Indesit Company

Decio Giulio Riccardo Carugati

Indesit Company

The evolution of the species

Electa

Cover
Prototype by Giorgetto Giugiaro
for Indesit Smart & Simple, detail

On pages 8-9
Hieronymus Bosch, *Paradise and Hell: Ascent of the Blessed into Paradise*, detail, 1500-1504. Venice, Palazzo Ducale

Graphic Design
Dario Tagliabue

Page Layout
Elisa Seghezzi

Editorial Coordinator
Caterina Giavotto

Editing
Paola Noé

Photograph Research
Arianna Coassolo

Technical Coordinator
Andrea Panozzo

Quality Control
Giancarlo Berti

Translation
Richard Sadleir

www.electaweb.it

"...So, by the mere fact of existing and hence being assigned to some function—explained Augusto Morello in his paper at the Beda conference in Brussels—objects inform about the services inherent in the culture of the society, in which they are employed and produced...." The assets described as queen goods, those which "actually acquire the dignity of paradigms of important changes in the history of humanity, or, at least, of different cultures", include home appliances. "Objects pass from lesser to greater complexity; they compete against each other, so determining a phylogenesis that goes from the abstract to the concrete, meaning they often foster the construction of a system suited to their efficacious and efficient use."
In this respect Indesit Company embodies a paradigmatic evolution, where the "species", the assets produced, are a project within the project.
The four "stanzas" of History, Design and Evolution, Sociology, Communication—the icons like musical tempos cadence the impact—investigate the phenomenon as autonomous compositions. The authors, in order of appearance, are: Giuseppe Turani, Enrico Morteo, Maria Sebregondi, Beba Marsano. I wish to thank Simona Giorgetti for her helpfulness in securing me the necessary contacts, Lea Ricciardi, her assistant, for the patient compilation of the iconography, extremely representative, in the order given, of qualitative dialectic of Indesit Company.
I wish to thank Dario Tagliabue for his handling of the graphic design, Caterina Giavotto, Elisa Seghezzi, Arianna Coassolo, Paola Noé, Giancarlo Berti and, in the person of Stefano Peccatori, all the editorial staff at Electa.

Decio Giulio Riccardo Carugati

Contents

History

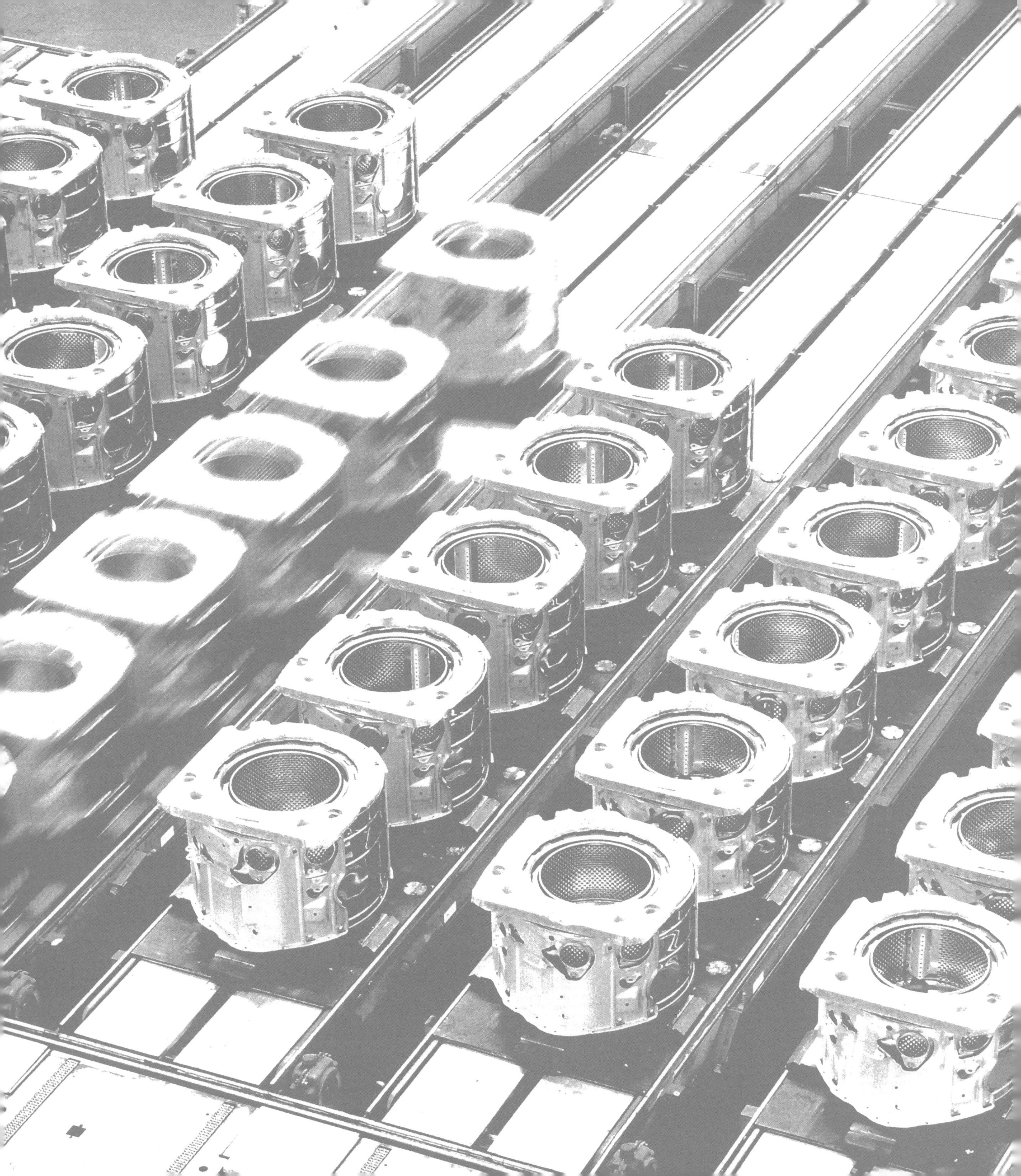

Giuseppe Turani

Some years ago, while I was at dinner in a restaurant in Rome with Vittorio Merloni, I coined a phrase for Indesit Company: "pocket multinational." It later proved something of a success and was even used to categorize certain Italian firms, not big but definitely present on foreign markets, strongly "internationalized", as they say nowadays. In effect, the Company was already then a multinational and it has become even more of one, eventually representing a sort of model in the Italian industrial panorama of recent years. This adventure is full of interest, since Merloni began as a modest manufacturer of scales and weighing machines, set up by an imaginative entrepreneur (Vittorio's father Aristide) at Fabriano, a small town in the Italian Marche. If there's a case of evolution in Italian industry it's precisely this: how to transform itself from a small business in the province into a real multinational, today with bigger investments abroad than in Italy. And the evolution finds its form in the moment when, in January 2005, Merloni Elettrodomestici changes its name to Indesit Company.

A pocket multinational

To understand immediately what we're talking about there's no better way than to have a look at Merloni's accounts for 2003. This will bring some surprises. The first thing to notice is the multinational dimension of the society, but probably what will really strike you is its size, and still more its growth in foreign countries. But let's look at the numbers. In 2003 Merloni produced just under 13 million appliances. This makes it the third-biggest European producer, after the giants Electrolux and Bosch (but in washing-machines it is second only to Electrolux). And this in itself is no mean achievement, if you think that the firm was set up in the thirties with 50 employees to make scales and weighing machines.

Add to this that in the meantime a lot of Italian manufacturers of home appliances, after an initial boom, have hit hard times and been taken over. They include two real market leaders like Zanussi (which fell into the hands of Electrolux) and Ignis (taken over by the Dutch firm Philips). In short, here we have a success story set in a place and time that isn't particularly notable for its successes. The Italian home-appliance sector had an extraordinary flowering in the fifties and sixties, and then in the seventies it went through a long dry patch: a lot of firms closed down and others were sold off. Indesit Company swam against the tide. And we can say still more. Its success would have been even more rapid and impressive if, when Zanussi was in difficulties, it had not been sold abroad. Merloni, like other manufacturers, made a bid for it, but Zanussi's ownership preferred Electrolux. Yet despite this, Merloni succeeded in following its own path and becoming third-biggest in Europe, right behind Electrolux, which had taken over the Italian Zanussi in the seventies.

We can add some details, without any fear of revealing secrets. In the eighties, when the group was shaping up well but still in the middle of the river (meaning it had not yet achieved European dimensions), it received a lot of offers from abroad: people who wanted to buy the whole business and on very good terms. And in those years the competition seemed really tough, if not actually unbeatable (as some consultants were heard to observe). Merloni Elettrodomestici's shareholders, partly from business

pride, partly from calculation (perhaps) finally decided not to sell and twenty years later it was clear they were right. To return to the group's balance sheet: in reality, behind the numbers there lies a lot else. There is a drive towards internationalization that is so strong as to seem almost exaggerated. And this expansion outside Italy also goes rather against the general trend, just when Italian businesses are having increasing difficulty in making headway on foreign markets. To put it clearly and explain what has happened, here, again, are the figures. Of 13 million home appliances manufactured by the Merloni group in 2003, just under 7.5 million were made in factories in Italy; the other 5.5 million were made abroad. It is perfectly clear (since the firm is planning to create or extend its foreign operations) that within a few years the group's output abroad will reach, and perhaps overtake, its output in Italy. But if the figures for output record real internationalization (and not just sales, meaning by commercial agencies, but actual factory output, after-sales networks, etc.), those for employment are even more explicit. Today the group, paradoxically, is stronger abroad than at home. The dynamics of new hirings are really impressive. In this case it is possible (thanks to data elaborated by Mediobanca's R&D department) to examine a period that's recent but fairly extensive: from 1999 to 2003. In 1999 the Merloni group had a labor force of just over 5,000 in Italy and just under 2,200 abroad.
In 2003, four years later, the force in Italy remained much the same (600 units more), while those abroad had grown from 2,191 to over 13,000. In substance, today the group employs just over 20,000 people, but more than 13,000 of them actually work in factories abroad.
At this point, looking at these figures, you might get a suspicion that this is a simple case of relocation: manufacturing abroad because it costs less. And probably this is part of the story, but if we continue our analysis, we see that it's far from the whole story. In 1999 the total turnover of the group was €1,420 million. Of this about one third (€409 million) represented sales in Italy. Already in 1999, that is to say, two-thirds of the group's output was sold abroad. By 2003 this proportion had been changed even further and all in favor of foreign sales. Put briefly, in 2003 Merloni's turnover was just over 3 billion, but the value of sales in Italy remained substantially the same as in 1999: at that time they had been €409 million, but by 2003 they had increased to €468 million. But in the same period Merloni's total turnover more than doubled.
We can say roughly that, while in 1999 the group's Italian sales were about one third of total turnover, in 2003 they were one sixth. All the growth (doubling the turnover between 1999 and 2003) came from expansion of sales abroad. This expansion was also accompanied by an expansion of production facilities abroad. So this is not just a simple case of relocation but a business that understands that its future lies abroad (on the European market, where half its production goes). And the company is tooling up to take both sales and production to those markets.
Putting it even more simply (and this is the extraordinary side of Indesit Company's evolution of the species), here

Scales and weighing machines first produced in the 1930s.

Grain from United States is loaded in Civitavecchia port, 1946.

we have a business that understood almost immediately it had to do two things: to grow fast and turn itself from an Italian business into a European one.

But we are also faced with a small mystery. In reality we have a family firm based in Fabriano that, within a few years, has succeeded not only in expanding strongly abroad but has also shown itself capable of running big production facilities (the one in Russia turns out 2 million appliances a year, and is the largest for home appliances in Europe). What's more, the same business has shown in these years that it is capable of running sales and service networks practically right across Europe. All this has achieved in a very short time, essentially a couple of decades. And, I might add, without the help of outside capital or bringing in some larger and more experienced group. In short, Merloni has not just become a pocket multinational, but has done all this "at home", with its own resources. If this is not the only case in Italy, it is definitely highly unusual.

Distribution of aid from Americans and "the train of friendship", 1948.

Manufacturing in the 1940s.

Opposite page
The factory, manufacturing and machinery in the 1940s.

The factory, molding department.

Growth

We have seen that in just four years the group succeeded in doubling its turnover (and all abroad), but this is only part of the story. Just to remember how it all began, we can cite some figures from one of the first profiles (perhaps the very first) that Mediobanca's R&D department devoted to Merloni Elettrodomestici. It dates from 1978 (so the figures refer to the 1977 balance sheet, almost thirty years ago). This was two years after the company was registered so as to give order and structure to initiatives that had gradually developed out of the Aristide Meroni's original business. By this time he had added the production of liquid-gas cylinders to scales and weighing machines. The profile is the portrait of a typical small to medium-sized Italian company. And, at that point, it was certainly difficult to bet much on this business. Its probable destiny, anyone would have guessed, was to be bought up by some larger and more competent group. Instead, as we have seen, things never turned out like that. But to return to the R&D profile of 1978, at that time the firm did not have a single factory abroad: all the production was based in Italy. Total output was less than two million pieces a year (including sanitary ware, which was later hived of in a separate company) against the present 13 million pieces. The work force numbered fewer than 3,000 (against about 20,000 at present). Sales in 1977 totaled 122 billion liras, of which, however, 50 billion liras were already represented by sales abroad, against the €3 billion today (about 6 thousand

billion liras in the old currency: even allowing for the fall in the value of the lira since then, it remains a big leap). Here is probably one of the keys to its future transformation. Already in 1977, when it was still a medium-sized all-Italian business, it exported 40 per cent of its output to foreign markets. Already then, in substance, Merloni had made an important discovery: its "natural" market was in Europe. Confining itself to the Italian market would mean dying of suffocation. And the "leap" from an Italian to a European scale, with production facilities stretching from Turkey to Russia, was far from easy. It took at least a couple of decades to complete. But in the end the operation was a success.

Something remains to be added to this brief portrait. In that period Merloni really was a typical Italian business. The factories were few, small and nearly all concentrated in the valleys around Fabriano. Often the Merloni factory was the sole factory in a given area and this clearly facilitated relations with the workforce and the social environment. In fact there are some who have spoken of this production setup as the "Merloni model": small factories in small towns, peaceful communities and high productivity. In those days things really were like that. In fact it is clear that in the first phase of Merloni's life it really was the "Merloni model" that sheltered the business from the crisis that struck (and sunk) other groups with a far better title to success. Indesit Company (thanks to its philosophy of "small and secluded is beautiful") managed to keep going. But once again a curious feature of this story emerges. Later, when the company began to expand and factories were built across Europe, they were no longer small and naturally called for the kind of management appropriate to a multinational. What is again surprising is how the people at Indesit Company succeeded, almost without changing stride, in passing from a domestic model (in the sense that everything was located just a few miles from home) to a European one. The only possible explanation, that probably also holds good for all the other factors in this story, is that at Merloni, in its DNA, there was a predisposition to successful evolution of the species. As we know, evoluiton of the species, also in nature, is never an easy and peaceful exercise. Often it leads to a dead end and instead of evolving the gene disappears. Indesit Company, clearly, had a rather special

Transport and manufacture of liquid-gas cylinders, 1935-1940.

Alcide De Gasperi, accompanied by Enrico Mattei, opens the degassing station at Cortemaggiore, 1952.

DNA. Within a couple of decades it became the opposite of what it had been, and all this instead of weakening it, instead of absorbing energies and shackling it, made it stronger. Why? We will deal with this further on, but here we can anticipate that, though small, the Merloni of 1978 (the little firm whose factories were all in the valleys around Fabriano), had two things:
1- The urge to be big one day, though perhaps at that time they still didn't know how.
2- The awareness that "that Merloni" was the first kind it seemed natural to create, but also the awareness that to become really big it had to take a big leap forward. To find another culture, different models of organization, different motivations. And in fact there were already authoritative consultants (some with international reputations), influential economists, experts in marketing and advertising at work on "that Merloni".
In short, the small company of 1978 (the one that appears in Mediobanca's R&D breakdown) recalls certain small-town intellectuals who hang out all day at the café round the corner but go to see all the foreign films, subscribe to magazines published in France and America, and make a point of talking to all the foreigners passing through town. This image may seem irreverent, but I believe it's right.
The small Merloni Elettrodomestici of the late seventies was a company just itching to be big, important, and step into a different suit of clothes. So it traveled the world, listened, learnt, studied (by the Harvard Business School, not by the accountant round the corner), and it took a lot of notes. True, it was still small, but it was studying how to grow big. The miracle is that when the opportunities occurred, the business (as if guided by its good angel) never made a wrong move. It made the right purchases at the right time, went public when it was right to go public, opened factories abroad when the time was ripe, though each step was rightly seen as hedged round with difficulties and called for prudence and insight. Merloni, as we shall see later, expanded partly through acquisitions, but it did not go at it head down (as others did, carried away by success), buying everything within reach.
It did things calmly, took careful aim and has never made a bad purchase.

Warehouse for the liquid-gas cylinders in the 1950s.

Evolution of the species

Even in the company, when they talk about their history and the road taken, they prefer to concentrate on what happened after 1987, when the company made its IPO on the stock exchange. In one sense that could be called the start of the group's real, public, important history.
Instead it's interesting to remember something of the ear-

lier period, that gray area between 1975 and 1987. It's worth remembering the part of the history that is known almost exclusively to insiders at the company or, at most, in the sector. In those years the company continued to make appliances and sell them: it was business as usual. But some features need to be stressed. It may not seem to differ from any other medium-sized Italian companies. But appearances are deceptive. Right from the start, even before the expansion began, it had some very original features.

Design and technology

The first of these was certainly the attention the firm devoted to technology and design. Nothing overpowering: at Fabriano (the company headquarters) they knew well enough how to turn out products that had to be "popular" and so could hardly indulge in freaks and novelties. However they also knew that, within the limits imposed by the market, they had to try to make things that were functional, up-to-date and attractive. They also experimented, turning out some very beautiful product lines that never went into regular production. I can still remember Vittorio Merloni in those years going to the traditional Cologne trade fair to exhibit his products. Naturally his whole staff went along too, because it was an important chance to make contacts, feel the pulse of the market and see what the competition was doing.

But on the little plane (that's what we called it, but it was a Citation business jet: and here perhaps it's worth recalling that Vittorio Merloni shared his first jet with another businessman who went a long way, Silvio Berlusconi) – anyway aboard the plane headed for the Cologne trade fair, together with Vittorio Merloni there were two passengers who had nothing to do with fridges and dishwashers, namely Professor Francesco Alberoni and yours truly. This was strange in itself, our interest in the biggest European home appliance fair was obviously zero. But Vittorio Merloni had a plan, and two objectives. At that fair he was presenting a line of appliances that was rather different from the firm's normal production, a very upmarket line with explicit design appeal. He wanted us to see it without explaining anything beforehand, to see what effect it would have. But then he also wanted us to have a look round to see how Merloni's products compared directly with those from other manufacturers.

Neither Alberoni nor I were experts in marketing industrial products. But that hardly mattered. Merloni saw us as two careful observers from outside the sector, and he wanted to measure with different eyes from his own and his staff's the value, the weight, of what he was doing. I believe he did the same with others. In short, while he worked he was continuously measuring himself, looking at himself in the mirror, seeing himself with different eyes, trying to see if he was on the right road or going astray.

And this is a regular characteristic of Vittorio Merloni. He will take you to look at his new factory in Russia (a colossal investment with colossal expenditure) around which

Opposite page
Visit by Alcide De Gasperi, Enrico Mattei, and Ezio Vanoni at Cortemaggiore in the early fifties.

Aristide Merloni among the crowd, 1955.

Trieste returns to Italy: the Italian army welcomed by an enthusiastic crowd, 1954.

he pondered, did his sums and consulted experts for months, but then he would want to know your opinion, would listen to it, take note and if, by chance, you suggested something that could be improved, he would make sure it got done.

I do not say that the "mystery" of the firm's DNA (and that does to confront them an evolution of the species important, but completed successfully) is actually this continual measuring himself and questioning himself and, perhaps, even more his ability to act as if he was bigger than he really was played an important part. In short, the firm's dimensions at that time were not decisive. The secret lay in being ready, in absorbing culture. Though small he sought to behave the way he would once the firm become big. So his consultants were all first-rate (international if possible), the designers were the best on the market, the experts were chosen from those that counted. And he traveled widely, to see his own factories, his competitors', and also those that were known for being attractive and efficient even if they made other things.

If this was true of design, it was truer still of technology. Here there were never delays. He realized from the first that home appliances (seemingly downmarket products) had an important technological evolution of their own. And so he never failed to focus on this factor. The result in recent years was the construction of some of the most technologically advanced production facilities in the world, with a truly impressive level of automation.

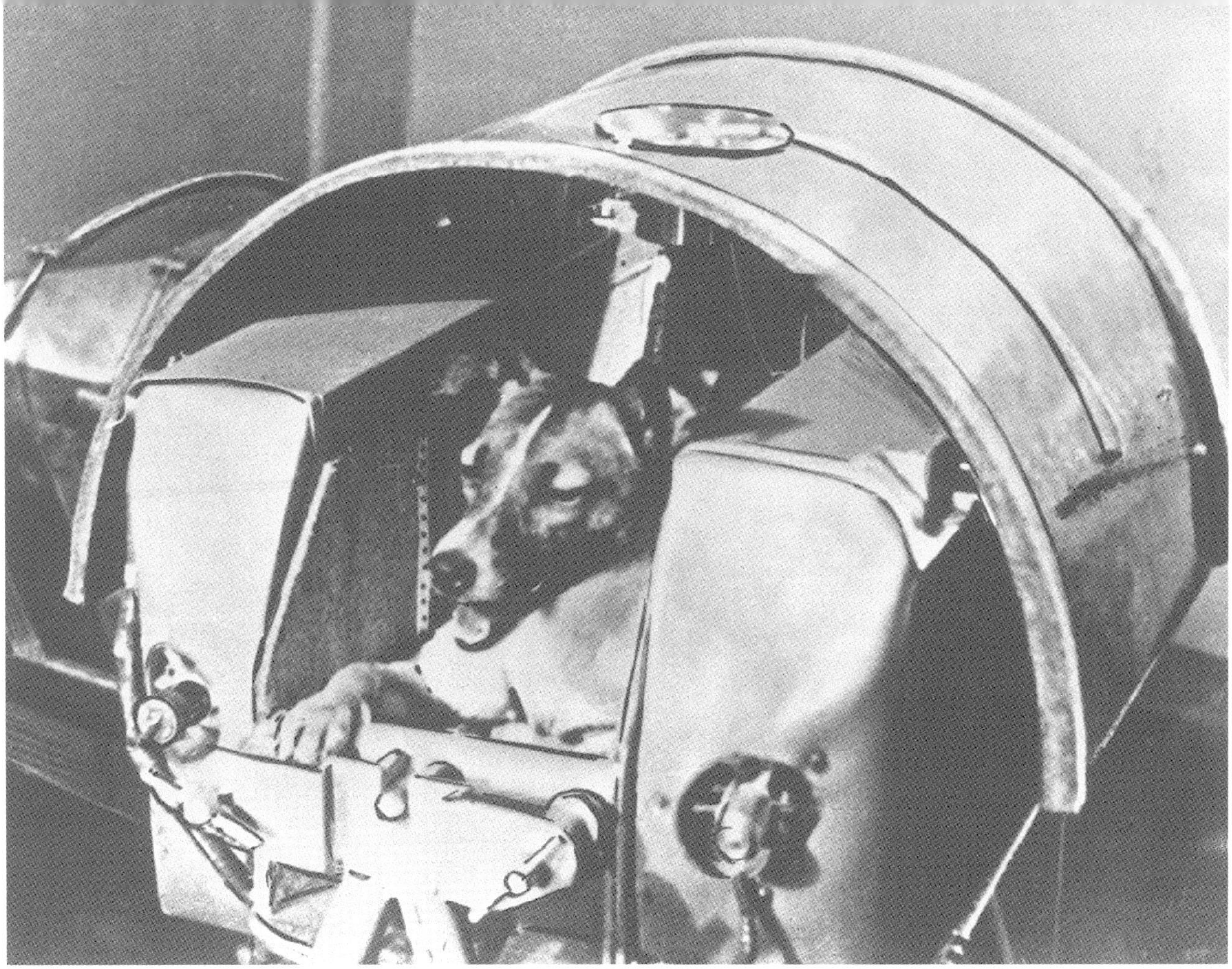

Space-dog Laika
on a trip to the space.

Here the direction was always clear. In the case of washing machines he understood that, even if the technology can't deliver yet, the goal has to be a situation where the user throws all the clothes into the machine unsorted and it just "knows" what it has to do with each without anyone pressing a button. The same holds good (in different ways) for refrigerators,dishwashers, and ovens.
I cannot tell, frankly, what point we've reached. But I know that at Indesit Company the road clearly leads in this direction. And I also know that it won't be an easy road to travel. It means continually monitoring all the relevant technology (sensors, dedicated chips, software, etc.) And the technology may come from America, but also from Japan or Korea. And so it is necessary to have eyes practically everywhere. And be ready to understand when a technology is finally ripe for installing in an appliance.
Another example is appliances linked by Internet. And here there are at least two evolutionary roads (I believe for the moment still at the initial stage). The first is the production of appliances linked via Internet with the final objective of making, for example, washing machines that are no longer sold to the customer but only rented) and whose use is then explained via Internet. And Internet also delivers instructions, new software, repairs, updates, etc. The second line of development is the intelligent refrigerator, that knows what is put inside it, that knows when things (for example milk) are about to run out and automatically orders them from the supermarket (or perhaps more modestly reminds the householder that there are only two pieces of prime steak left). These are only examples (and, I repeat, I cannot say how close we are to achieving them), but they serve to explain that for years at Fabriano they have no longer been thinking of a refrigerator as a simple "machine for producing cold." Appliances are seen as "machines" within the broader revolution in information technology of recent years. A revolution that has already altered automobiles, for example, and had some important effects on home appliances (and will have still more in the years to come). In short, we can close this chapter by saying that certainly Merloni packed into all its appliances the technology that could be packed into them. Instead of contenting themselves with remaking the same things over and over but spending less, the firm spent more to continually update its products. In this sense the group has never been managed thriftily, but always looked a little further ahead. But "looking ahead" is not just a question of technology.

Ariston logo, 1960.

The Kennedy family in the early sixties.

In reality, it is a philosophy of life or, if you like, an attitude, an almost instinctive form of behavior. In this respect I can quote the words of Vittorio Merloni himself (under whose leadership this evolution of species took place). These words help explain all the progress the firm has made in such a short time (and also the mystery of the evolution of species that has taken place). We were in London, where Merloni had a production plant (not large) and a sales center, which for years had never turned a cent's profit. And obviously this seemed to be nonsense. Why go on? Vittorio Merloni's answer was very simple: "Here in England you have all the big manufacturers of domestic appliances from around the world, so it is a very competitive market, perhaps the most competitive there is. We don't make money, but we learn a heck of a lot, because we see how the others are doing things, at firms that are bigger than and sometimes better than us. But above all if there's a new product development (or a new approach to sales and service) we'll see it here immediately."
In short, instead of being satisfied to tend its own patch, its home market, Indesit Company took the opposite road: it made comparisons with the big firms so that one day it would be "like them." And seeing it is now third-biggest in the sector in Europe, this logic, this policy, has certainly paid off.

Looking out of doors

The firm has looked out of doors over the years in a number of ways. Firstly by simply travelling and talking to anyone who had something interesting to say, whether they were technocrats or marketing experts.
Then there was and is perhaps the most important instrument: Merloni Progetti. As the name suggests, this is a company set up to design factories all round the globe. It has enabled the group to build up contacts (and friendships) in many parts of the world (from Russia to China) and to keep its finger on the pulse of the situation abroad most of the time. In addition, at least ever since the IPO, there have been

continuous contacts with major international bankers for acquisitions, joint-ventures and agreements of various kinds. And here it is important to stress than in international financial circles, the corporation has been seen for many years as a group in growth and therefore potentially interested in moving into different countries. So when some "bargain" comes onto the market for appliances (say a company for sale in some emerging country), Indesit Company is one of the first to be approached.

things everyone does on vacation, above all in such a beautiful spot. We went out boating, we sunbathed, we swam and in the evening we went looking for a good restaurant. Vittorio Merloni had just made his offer to take over Indesit. But he was far from confident, afraid he had offered too little. The thought kept nagging at him, to the point where, repeating, "I don't want to bother you," he began to tell me all about Indesit. The whys and the wherefores, his misgivings, the reasons for and against. Eventually he came

Vittorio Merloni with Deng Xiao Ping, Giovanni Agnelli, and Leopoldo Pirelli in China, 1975.

Opposite page
Company's registration to Chamber of Commerce, 1975.

Pa-Vision Award won for social responsibility, in 1974.

Investments

In the group's growth there are at least two important secrets. The first is certainly the urge to grow. Or rather the awareness that it has to grow if it is not to lag behind. This feeling was already clear years back, when Merloni was still a small group, and it's even clearer now it is one of the big players in Europe and worldwide. But it's one thing to be generally willing to grow and another to have the courage, the ability, to make the leap when the moment comes.
In this respect there's the very significant story of the acquisition of Indesit (which was to change the history of Merloni Elettrodomestici by transforming it into Indesit Company). I still remember the Indesit story. It was in 1985 in Vittorio Merloni's house in Sardinia, right on the seafront. We were on vacation and so we were doing the

to a conclusion: "If I take it, and if goes well for me, it'll be a big leap forward. A leap that would take me years and years to achieve with my own resources, beefing up sales and production. But if someone else takes it, then they'll get a head start and perhaps I'll never be able to catch up. Well, either I jump or I don't.
Or rather, I have to jump now or Il have to pull out of the race, sell everything."
In the end he went to the phone and raised his offer substantially, so succeeding in winning Indesit. Mulling over the business like this (should I or shouldn't I?) may seem surprising, but there is an explanation, a very serious one. The outcome of this acquisition explains clearly the spirit of the company. Indesit was not just any company. It was Merloni Elettrodomestici's strongest competitor and it

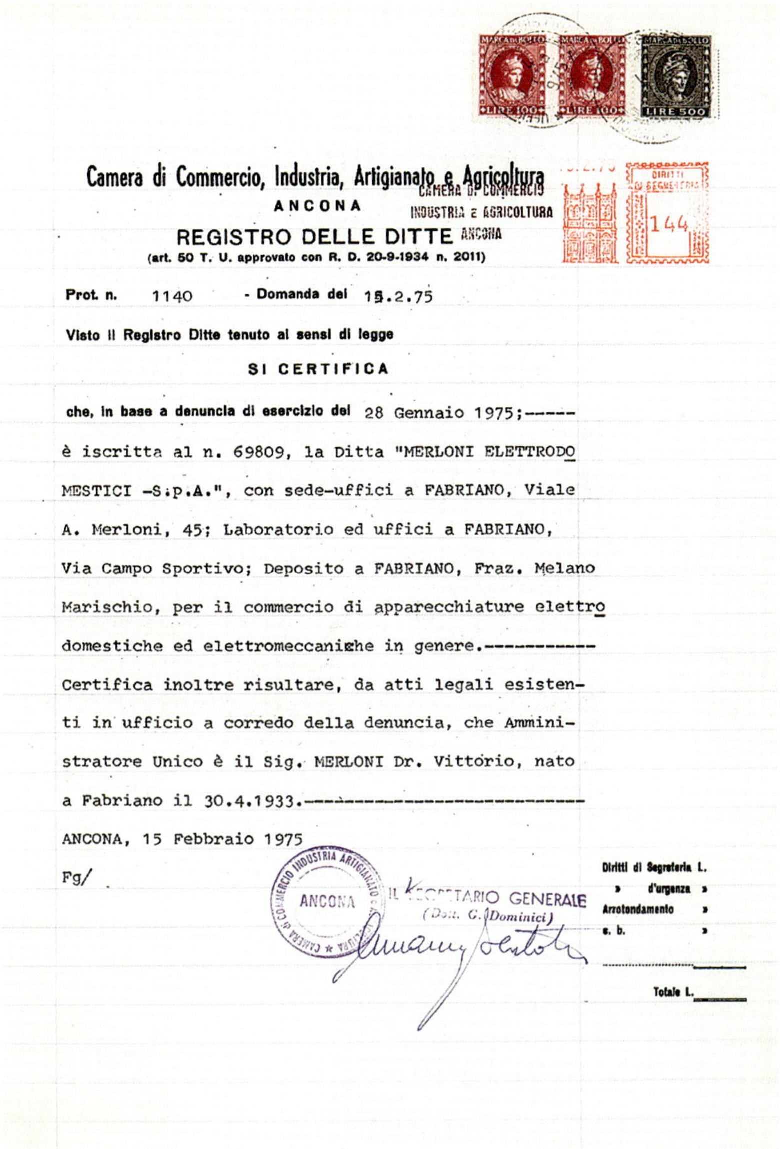
Camera di Commercio, Industria, Artigianato e Agricoltura
ANCONA
REGISTRO DELLE DITTE
(art. 50 T. U. approvato con R. D. 20-9-1934 n. 2011)

CAMERA DI COMMERCIO INDUSTRIA E AGRICOLTURA ANCONA

DIRITTI DI SEGRETERIA 144

Prot. n. 1140 - Domanda del 15.2.75

Visto il Registro Ditte tenuto ai sensi di legge

SI CERTIFICA

che, in base a denuncia di esercizio del 28 Gennaio 1975;-----
è iscritta al n. 69809, la Ditta "MERLONI ELETTRODOMESTICI -S.p.A.", con sede-uffici a FABRIANO, Viale A. Merloni, 45; Laboratorio ed uffici a FABRIANO, Via Campo Sportivo; Deposito a FABRIANO, Fraz. Melano Marischio, per il commercio di apparecchiature elettrodomestiche ed elettromeccaniche in genere.-----------
Certifica inoltre risultare, da atti legali esistenti in ufficio a corredo della denuncia, che Amministratore Unico è il Sig. MERLONI Dr. Vittorio, nato a Fabriano il 30.4.1933.-------------------------

ANCONA, 15 Febbraio 1975

Fg/

IL SEGRETARIO GENERALE
(Dott. G. Dominici)

Diritti di Segreteria L.
» d'urgenza »
Arrotondamento »
s. b. »
Totale L.

was much stronger industrially (even though it was floundering). Indesit, which had been founded in 1975 together with Merloni but with a company capital ten times larger, also had an extensive network in Europe.
Though not actually a giant, Indesit was a very big mouthful compared with Merloni. To the company based in Fabriano, it might well have seemed, as they say, that it was biting off more than it could chew by acquiring a company of that size. But on the other hand the acquisition of Indesit, at bottom, was almost its only chance to make an important leap into Europe, to gain at once a position that otherwise would have taken years and years of hard work. And so the operation was carried out, with some trepidation at first. And, if I am right about the feeling in those days, also with some forebodings, well aware that it was making a leap in the dark. Merloni Elettrodomestici had never made such a big acquisition before, it had never tried to take over a business so strongly structured, with such a strong identity and image. In short, Indesit was not a business stuck in the provinces and happy to be absorbed by a larger Merloni Elettrodomestici. In a certain sense almost the opposite was true. And in fact I remember clearly that some months later Vittorio Merloni was grappling with a problem that was absolutely new to him and his staff. He had discovered that Indesit had a business culture of its own, meaning its own way of working, and Merloni Elettrodomestici had different one. Between the two cultures very serious conflicts sprang up, making it difficult to master the culture at Indesit. Dozens of meetings were spent chewing over the problem and I believe that in the

A youthful Makio Hasuike, at work.

end the people from Merloni Elettrodomestici had to insist they were the owners and so held the winning culture by definition. But the story does not end here. Indesit in a sense has made a comeback. In fact from January 2005 Merloni Elettrodomestici, as we know will be called Indesit Company. This is a fair-minded recognition, by the firm that was strong enough to buy "poor" Indesit when it was in trouble, that the Indesit brand was stronger in Europe.
Perhaps my conclusion is excessive, but I believe without that acquisition in 1985, without that act of courage (without that willingness to take a risk), today we would not be here talking about Indesit Company's success. And this is the proof of what I said above. It's not enough to be competent, to do research, to focus on costs and earnings. In the history of companies there comes a time when an act of courage is required, when a firm finds itself in a tight spot. And in 1985 the company had this courage. It showed the same courage later, when the opportunity came to take over the Russian factory at Lipetzk.
The second key to the firm's successful growth is investment. In practice 85 per cent of the firm's business cash flow goes on investment. The figure may seem very high, too high. But, they explain, this is the practice at the leading Japanese companies, which they have studied carefully (especially Toyota). And so they do the same at Fabriano, too. Further confirmation that the fact of being small (as this firm was years ago) never prevented it from acting big.

One single company

There is another fact that should be noted. So far we have seen that Indesit Company (known commercially all these years as Merloni Elettrodomestici and commercially combined with two Europe-wide brands Ariston, from the name of the founder, Aristide, and Indesit, as well as the regional brands Hotpoint, Scholtès and Stinol), has grown steadily and continuously. It has to be said that half this was internal growth (meaning expanding the business, increasing market share and building factories) while the other came from acquisitions in Italy and Europe.
And here, if we like, there's another small miracle. Despite this continuous growth and acquisitions in various parts of Europe, the company succeeded in never losing its identity, remaining substantially what it was at the start. You only need to have seen (many) other mergers and takeovers turn out badly, precisely because of differences

in their cultures, to realize that what has been achieved is not simple. I remember that the problem also cropped up in the company (as explained above), with serious repercussions when Indesit was absorbed into the group. Indesit was a big company and had a strong culture of its own. And so at first things were more difficult. In the end, it was decided that the values that should count and prevail were those of Merloni and those not of Indesit. And so it was. On the other hand the set of Merloni values has a long pedigree (it goes back to the founder, Aristide) and is also very simple. In the firm they have been summed up like this: innovation always, in products, in services, in the life of the consumer; respect for the environment, for sustainable development and transparency towards people; ambition for leadership to achieve goals and meet the shareholders' requirements; concern for and closeness to the needs of stakeholders; simplicity and genuineness.
Clearly these are simple principles and with nothing revolutionary about them. Besides, it is hard to see what more a manufacturer of home appliances could do, but, over these thirty years they have been applied strictly and this is what has enabled the company to grow fast and to incorporate a wide range of different businesses without losing its way. At Fabriano I saw in person one thing that probably explains more than any complicated arguments what it is, ultimately, that makes it "one single company." In the accounts department there are different desks, each of which has a different flag: Italy, France, Britain, Russia, etc. On these desks arrive the accounts and the requests from the various businesses of the group. And if a problem arises the account clerks speak to each other, swap data, unify the procedures and the replies to give to the various questions. In a sense, that room is "one single company." Basically, if the Russian company is looking for a certain type of switch and the British company has found a good supplier for that component the feedback is immediate and the contact is set up in a few seconds. But, above all, the different sets of accounts are unified (in criteria and methods) and so everything runs a lot more smoothly and efficiently. Some people might see all that as rather ingenuous. At bottom, today all this could be done using information technology (a field the company has plenty of expertise in) without keeping the clerks all in the same room. So perhaps one day the clerks in the room will stay at home and everything will travel along IT cables. But I should say that seeing all those desks with the flags

Young girl demonstrating for peace in the 1970s.

Prince Charles at his wedding to Diana Spencer, 1981.

Opposite page
Computerized station for quality control of washer-dryers at None (Turin), 1988.

Indesit logo, 1987.

Paint line for washing-machine casings at None (Turin), 1988.

of half Europe at Fabriano does make a certain impression. And it is clear that, if this was originally (like the whole Italian domestic appliance sector) a provincial firm, today things would no longer be done like this. But it is also clear that this is a company, a group, and not an aggregate of different businesses. In short, "one single company" is not a showy description but accurately describes a way of being and working. The business has grown fast, very fast, but it has not unraveled in the process. It has remained a body that moves with at the same rhythm and to the same tempo.

Going public

The group made its IPO in 1987. But the whole process had got underway well before. I would say there were two phases. In the first and very long one, there was discussion and reflection on the advisability of taking such a step. On the one hand, quotation on the stock exchange made it possible to draw on the financial markets and find bank finance more simply and directly. On the other hand it was felt it would create problems. The business would be under the eyes of everyone, its performance and services would be put under the magnifying glass by analysts.

At Fabriano they spent some months weighing the profs and cons. They saw the step was inevitable if the company was to become a big, mature corporation, and so it could hardly be put off endlessly. Moreover it was clear that to be accepted fully by the international financial community they had to make an IPO. But it was also felt that the step would not be simple. At bottom, till then Indesit Company had been a family business whose shareholders were descendants of the founder Aristide and no outsiders. On Christmas day, after lunch, they cast up the accounts and saw how things had gone.

In the end it was decided that the firm had to be quoted on the stock exchange because it was inconceivable it could continue to grow the way they wanted it to grow, becoming one of the first three or four manufacturers in

Assembly line for cookers at None (Turin), 1988.

Checking the esthetics of dishwashers at None (Turin), 1988.

Opposite page
The Space Shuttle Orbiter 101, Kennedy Space Centre (USA), 1980.

the sector in Europe, if it remained a family business. And I believe that in the end the decisive push came from this fact: the awareness that without a stock exchange quotation it was not a business with all the rights and duties that can be acquired on the international financial markets.

After this first phase, the second took off. Now it was a question of preparing the company to go public. Here there were two tasks to be done. The first, obviously, was to reorganize the accounts, passing from a "private" system, tailored to the needs of the Merloni family, to a more public system, that would satisfy the requirements of clarity and transparency demanded by the market and the supervisory bodies. In short, the firm had to comply with the requirements of the Milan stock exchange, the Consob and the other authorities in the presentation of its accounts. The process took some months. But in the end this step was also completed.

The second question was the organization of the shareholding within the family. This is not the place to dwell on this operation (very long and delicate). Suffice it to say that already in 1975 electrical appliances were hived off from thermo-sanitary products (which went to Francesco, substantially) while appliances went to Vittorio. At first the IPO (after all the work it had cost) was rather disappointing. The market, clearly, did not believe that an electrical appliance manufacturer had a future or was worth gambling on. There were times when Vittorio Merloni actually thought seriously about making a public offer for his own firm so as to be able to stop it from being quoted. But luckily everything remained as it was. Within a few years the stock exchange discovered that Merloni shares were very attractive and so there were no more second thoughts.

Reflecting afterwards on that decision, it should be said it good was a good thing, for various reasons. On the one hand the business went public and so has become accustomed to the culture of transparency and the importance of performing well and producing good results. On the other, the fact of being quoted prevented conflicts arising (as happens in many family businesses) about who does what, who has the last say, etc. But the most important advantage, perhaps, of going public was that today Indesit Company is a business that publishes a regular balance sheet, monitored by the market, and so, when it has to carry out an operation (however big) it no longer has to worry about explaining who it is and what its accounts are

The factory at None (Turin).

Vittorio Merloni, 1981.

like. Everything is public and transparent. It is certainly difficult to appraise how much this has helped the group in its growth, but it has certainly not proved an obstacle. And, finally, it has accustomed the firm to seeing and appraising itself with standard criteria, immediately comparable to those of its competitors. This may seem to be a detail, but it is far from negligible.

The managers arrive

Though the company went public in 1987, ten years passed before another revolution arrived, that of the managers. Naturally the company has always had managers.

But most of the decisions rested on the shoulders of Vittorio Merloni, who directed the group and was (and is) also the biggest shareholder. But in 1997 it was decided that the transformation begun when it was quoted on the stock exchange should be continued through a small managerial revolution. In substance, it involved appointing a team of managers and reserving the title of president for Vittorio Merloni.

It was bigger and more important than just a division of labor. In a business that for over twenty years had been "all family", for the first time it introduced a form of separation between ownership and management. In substance, the ownership sets the objectives and strategies (in cooperation with the managers) and then the managers are supposed to toil to achieve the objectives set. And on their success will depend the rewards (or punishments). Also in this case there is nothing revolutionary, that's the way things get done in the business world. But it is also well known that carrying through revolutions of this kind in businesses that have always been of the family type is neither simple nor immediate. But in 1997 Merloni also adopted this business style, even Aristide's "old" manufacturer of scales and weighing machines (which had started off with 50 employees). There were also misgivings over

Vittorio Merloni
and Sandro Pertini,
Milan, 1984.

Vittorio Merloni
at Confindustria
with Giovanni Agnelli, 1984.

Opposite, bottom, and below
Fall of the Berlin Wall, 1989.

this change, as over going public. But today no one any longer doubts that it was the right decision to take.
The business is far more transparent and better managed. And this is not because it's the biggest shareholder (Vittorio Merloni) who appraises his own performance as CEO of the company, but because Vittorio Merloni (together with the other shareholders, big and small) that appraise the value of its management team.

Indesit and Vision
Since history often contains reserves of irony, the latest decision of is that, from January 2005, the group is no longer called Merloni Elettrodomestici but Indesit Company. The old established name is abandoned. It is replaced by the name of the company bought years ago and which for a certain period caused problems as efforts were made to integrate it with the rest of the group's business. Why? This is also part of the process of growth.
Now that it has become the third-biggest European manufacturer of electrical appliances and one of the twelve biggest in the world, Merloni needed of a company name that would identify it easily and rapidly on the markets.
In the end, it has chosen to stake everything on Indesit because Indesit proved to be the better known brand in Europe and also because it has connotations (youth, dynamism, etc.) that are what most interest the company today. It has taken thirty years from the foundation, but in the end the journey can be said to be almost over.

Vittorio Merloni with John Paul II during Pope's visit in the factory, 1991.

Opposite page
Giorgetto Giugiaro at work.

The factory at Lodz, Poland (manufacturing cookers).

On pages 38-39
Attack on the Twin Towers, New York, 11 September 2001.

On page 40
The washing-machine factory at Comunanza (Ascoli Piceno).

On page 41
Vittorio Merloni in a photo by B. Krieger, 2000.

The old manufacturer of scales and weighing machines is quoted on the stock exchange, it is run by its managers and has its own company name which will identify it everywhere.

Any other business? Yes. We still have to look ahead, at the future. We need to have what the technocrats running the company call a 'vision'. An idea of what the firm might become or at least of the market in which the firm lives and seeks to grow. The elements of Indesit Company's vision are very simple, to the point where it can be summed up in two points:

1- It is necessary to be aware that today the global market (not European, but global) for electrical appliances is 80 percent in the hands of just 12 companies (including Indesit Company). Half these companies are already Asian. So there has been a big process of concentration and a shift eastwards. This clearly means the company urgently needs to grow but also to create accords for cooperation and joint-ventures with other producers. While once the challenge that interested Indesit Company was in Europe (and this was fought and won), now another challenge is looming up. But this time it is global, worldwide. It is a far more complex challenge quantitatively and qualitatively and so it calls for new tools.

2- We also need to be aware that today you win not just by producing a good product but also by supplying good after-sales service. Service is nothing new: the company has long run highly efficient call centers and, above all, it has several thousand service vans in Europe, all linked to the depots by radio. In the morning they all go out with a list of calls to make (based on the previous day's reports), and then instructions for urgent calls arrive by radio.

It's an impressive machine at the service of customers, but clearly a machine that will have to be further developed and refined. Here, as with the hardware, the real machines, progress is continuing. Precisely for this reason, to meet these different needs (the global challenge and the improvements offered to customers), various committees have been set up at the company. Two are observatories: the first one studies the American market and the second the Asian market, in quest of anything new that it could be useful and interesting to "import" into the company. In addition there is a committee that has the task of keeping track of technologies and innovations, searching for those that could be usefully introduced into the appliances, the new products. Finally, there is a committee that oversees the course of the financial stability plan and a service for human resources with international skills. And this is also a sign of the times. The company that once recruited its workers in the valleys around Fabriano now scours the world for technicians and specialists: the important thing is that they have to be smart and have good ideas.

The last challenge

So what does all this mean? Essentially that Indesit Company has come a long way in these thirty years, it has been very successful. And it has won a series of challenges. It has succeeded in being one of the Italian manufacturers of appliances that has survived and grown, and not one of those that have disappeared or been taken over.

And, to succeed in this, it had to undergo an evolution of species within itself. Then it has succeeded where few believed it could: it has won the European challenge, in

the sense that today it is on the short list of the three biggest businesses. At the same time it has taken to the limit its evolution of species from a family business based on Fabriano and its environs into a multinational, Europe-wide in scope, with almost 20,000 employees and a turnover of €3 billion.

But that was yesterday. Now we have come to the last challenge. Indesit Company is in the charmed circle of the 12 major companies that control the world market in appliances and it has to be capable of defending its position and growing further, even through partnerships and joint-ventures.

Can it win this last challenge? Seems so. As it has progressed, in fact, the company has greatly enriched itself. In addition, besides its old DNA of good values and the urge never to surrender, it has a lot going for it. Large size, the ability to place itself strategically and act simultaneously in a number of different countries, considerable mastery of the technologies and a knowledge of how to behave on the markets, and finally a company built according to the rules, with the managers on one side and the shareholders (large and small) on the other.

In short, the tools are all in place. Nothing is lacking.

And the challenge is just round the corner: or rather it has already begun.

Indesit Company

New Indesit Company factories.

On pages 44-45
Stefano Baldini triumphs at the Athens Olympics, 2004.

ATHENS 2004

ATHENS 2004
2157
ATHENS 2004

Design and Evolution

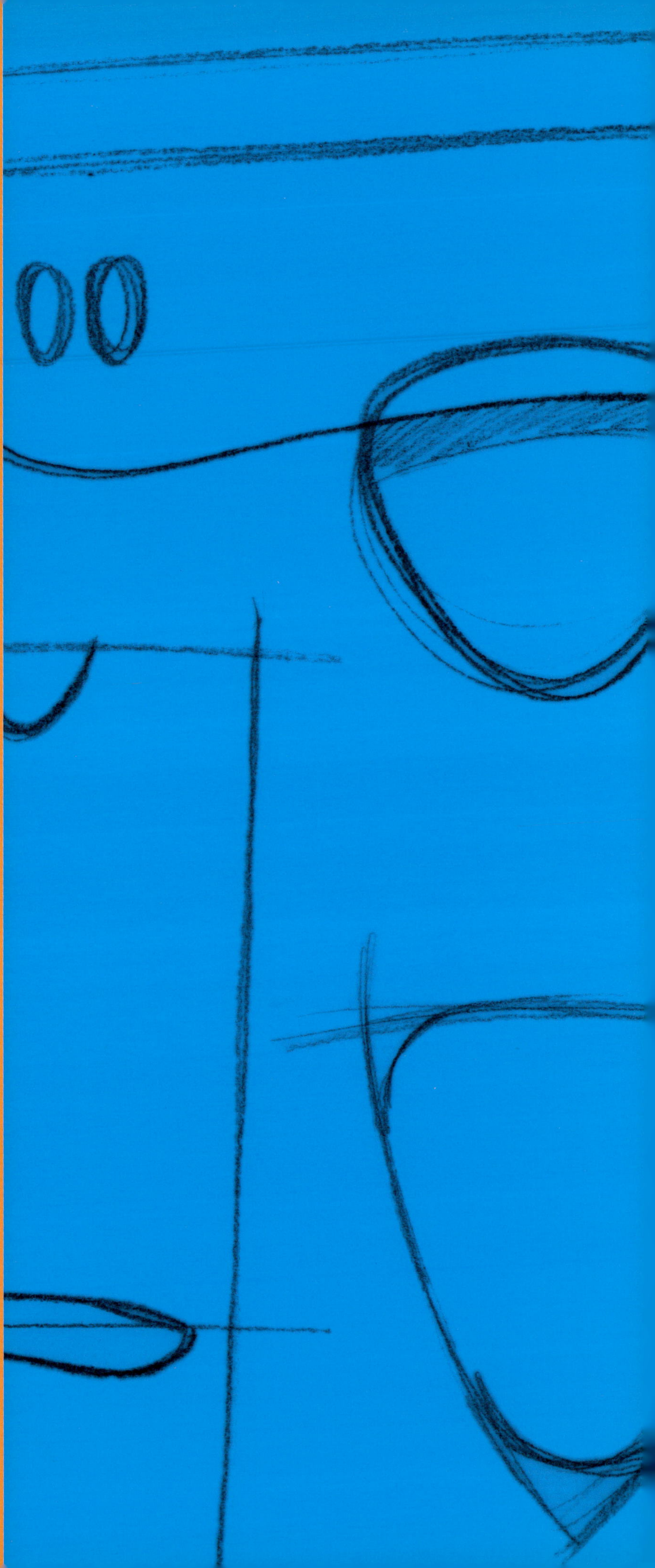

Enrico Morteo

Today washing machines, refrigerators and dishwashers are taken for granted, the commonplace inhabitants of our domestic landscape. But just a few decades ago they were still rare, the possession of a few privileged families. In the same way, their clear and linear forms, which now strike us as normal, almost natural, are the result of a typically Italian industrial project and design, which imposed great formal simplicity on both the increasing technological complexity and the evolving use of these supremely domestic appliances.
Essentially humdrum objects, home appliances reflected more than most other products Italy's industrial breakthrough in the postwar period. The rapid development the whole of the domestic appliance sector not only epitomizes the country's industrial expansion but also suggests the spirit of confident social optimism that fuelled the Italian design project, which was a not particularly hidden driving force behind the country's extraordinary boom years. Retracing, however briefly, some of the major landmarks of this development and finding traces of it in the personal vicissitudes of Aristide Merloni and the history of one of the most important industrial groups in Italy, will mean drawing a map of the profound connections that exist between production and project, shedding light on the roots of that design culture that is still one of the most vital resources of the Italian economy and a key feature of our industrial identity.

At the end of World War II the picture in Italy was contradictory, made up of contrasting lights and shades, but the signs of modernity were not sufficiently numerous to put its mark on the country's overall image. Left behind by the acceleration that was transforming the socio-economic makeup of the major European countries, Italy's production sector failed to achieve the immediate substantial changes in collective behavior dictated by the new industrial order. Fundamentally rural and peasant, Italy was still a country moving to agricultural rhythms. Significantly its most solid industrial sector long remained textiles, directly connected to the cultivation of cotton, but this was just the first step in an industrial development that grew slowly.
This does not mean it was totally lacking in all signs of modernity. Its liberal governments in the early decades of the twentieth century made significant advances in the provision of public infrastructures, encouraging shipbuilding, railroads and transport in general. Then there was a timid modernization of the public administration, which created new spaces and instruments for office work, as well as new buildings to house it. Concurrently, though under close political control, the banking system was strengthened to support private enterprise, in its turn called on commit itself to strategic areas of the development of a modern industrial system, such as steel-making and the production of energy. Despite the structural fragility of the national steel industry, which still weighs on the country's economic developments, Italian industry developed and produced excellent rolling stock for the railways, large and luxurious ocean liners, automobiles that were not only attractive but fast and victorious in road racing, revealing an up-to-date culture of design and manufacturing that, in small production runs, had no reason to envy the world's strongest economies.
The research center for rail transport in Florence, Savigliano

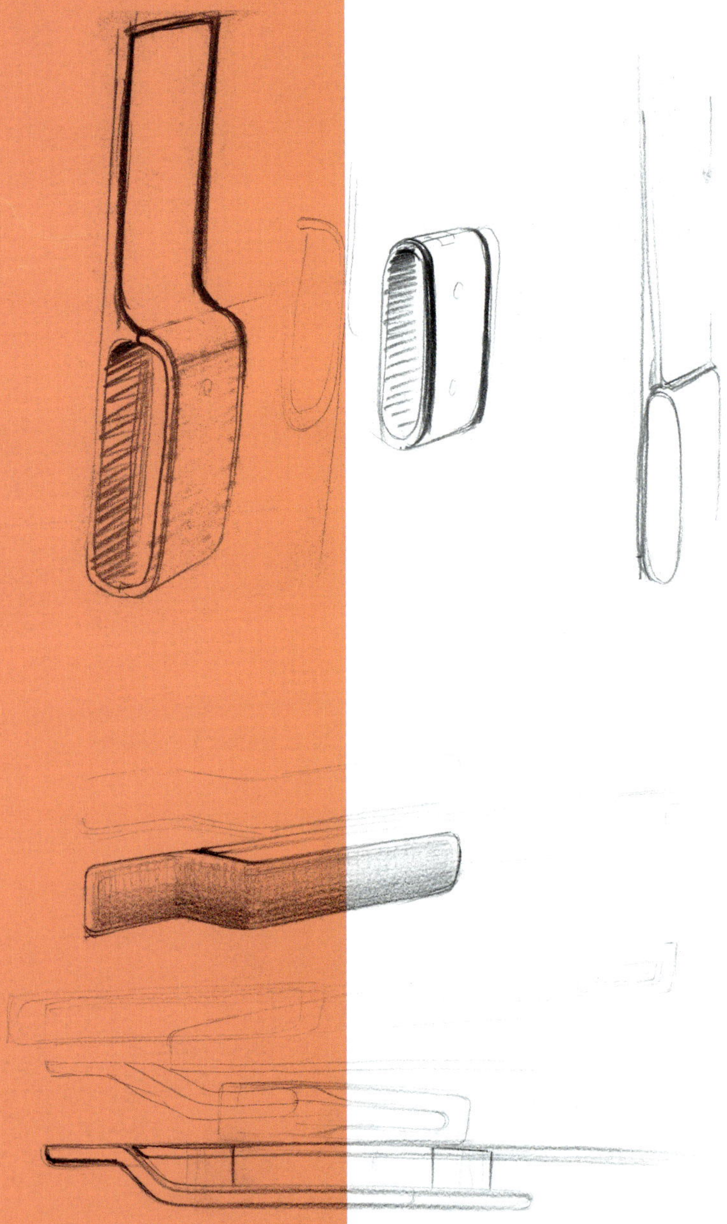

Makio Hasuike
and some sketches
for Ariston Unibloc.

Model and sketches by
Makio Hasuike for Ariston
Unibloc 5, refrigerator,
stainless-steel sink, rapid
water heater, dishwasher and
cooker, 1968.

Ferroviaria, Breda, Nuovo Pignone, Fiat, Olivetti, the shipyards in Genoa and Monfalcone: these were some of the numerous successes of Italian industry whose roots lay in that early season, out of which grew Italian manufacturing and design.
All the same, Italian industrial strategy remained orientated almost exclusively towards large-scale public orders, often military supplies, and only to a much smaller degree intended for the private consumer market. The same situation, through for different reasons, substantially continued under the Fascist system. Fearing the opposition of a strong urban grouping, both middle and working class, Mussolini always distrusted the possible development of large industrial concentrations in the cities, preferring a more diffuse policy of support for the rural world and subjecting industrial groups to rigid central control. Dalmine, the big steelworks, located in the countryside outside Bergamo, is a good example of this.
The substantial compression of the middle class and consequently of the market for private consumption did not, however, prevent the development of advanced sectors of research. On the contrary, the isolation dictated by the policy of "autarky" fuelled high rates of innovation and research in some fields. This was the case in the chemicals

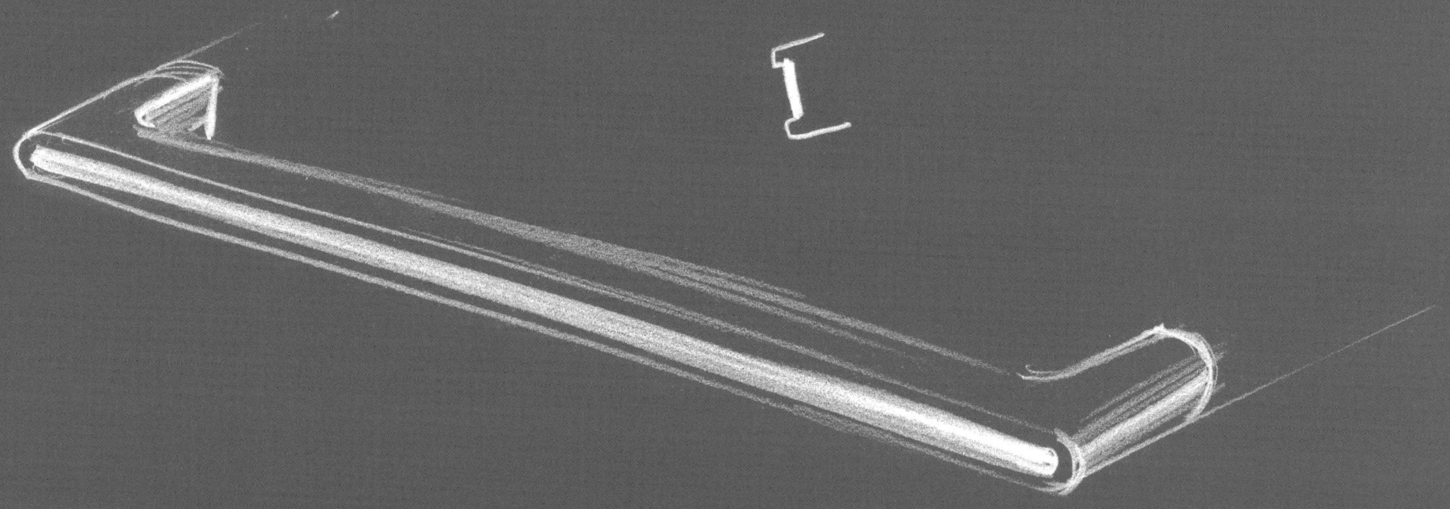

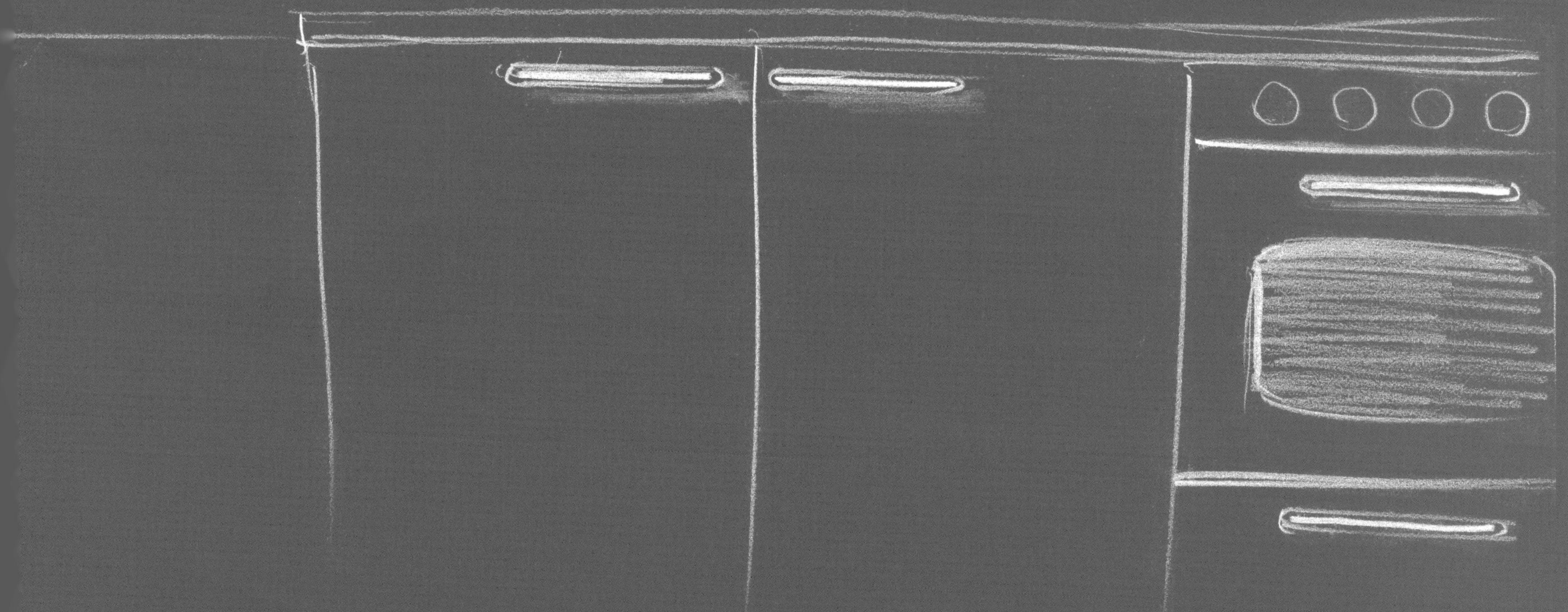

6 4.78 OSA

1. LB. cerniera 30/4

2. spia verde – ricerca materiale

=

~~20/4~~ ~~30/4~~

3. frontale forno elettronico

4 terminale verniciato

5 colonna

6. Scolapiatti

entro 30/5

7. lavelli

Tavolo

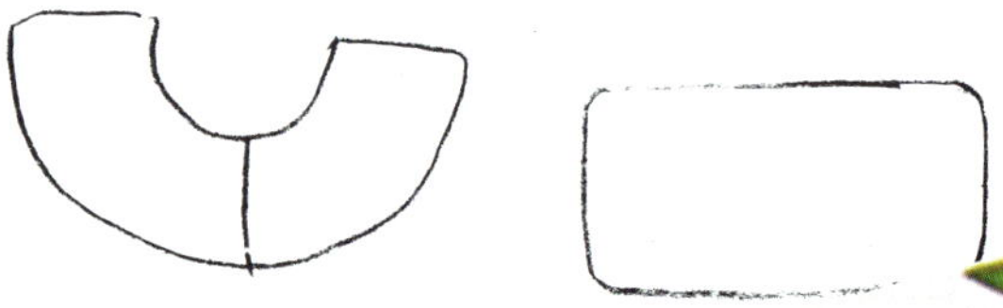

Sedie

incontro con Tecnova 13/4

Nome architetto

Model, sketches and notes
by Makio Hasuike
for Sistema OSA, 1977-1978.

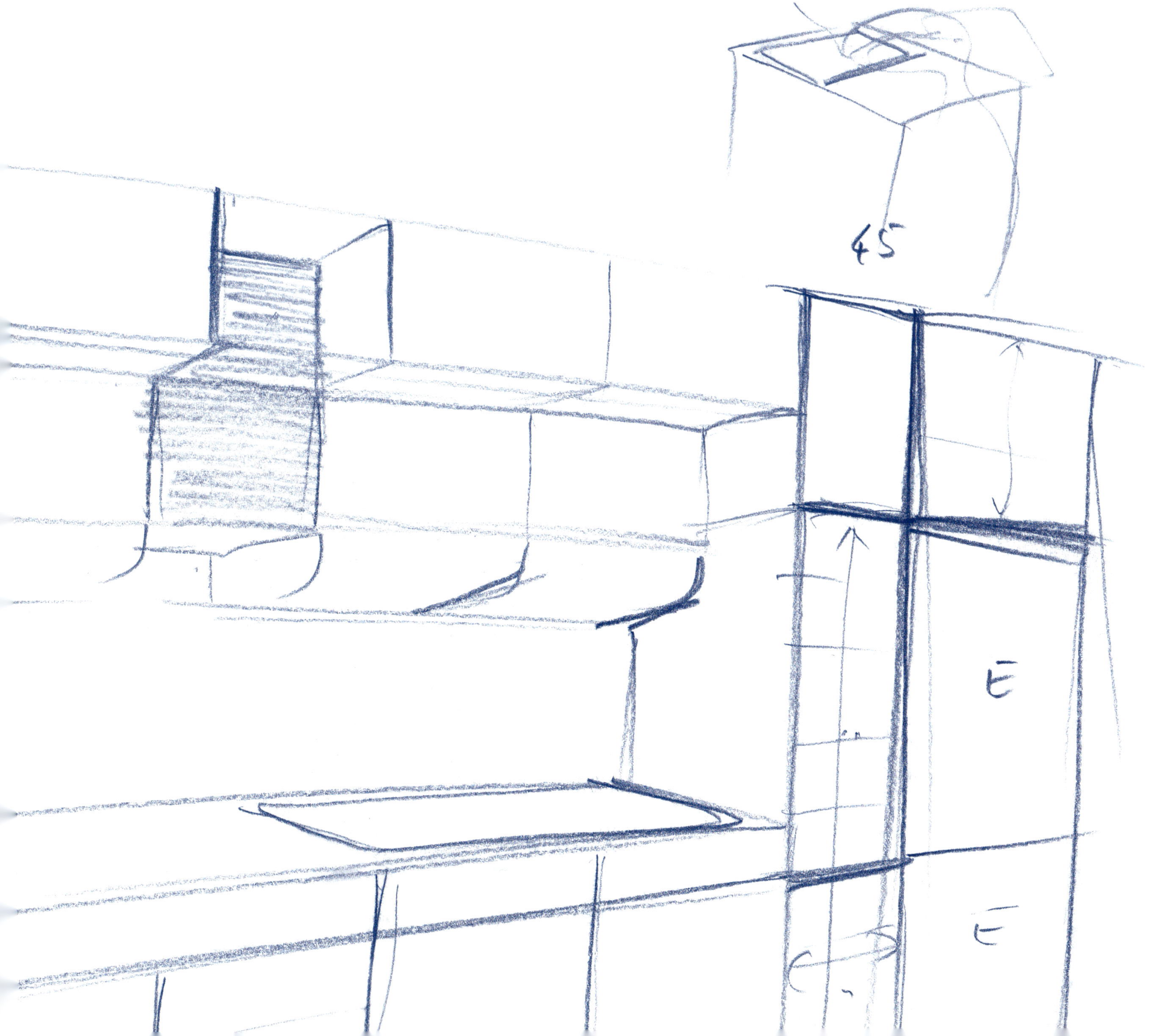

sector, where the shortage of petroleum spurred a search for alternative sources from which to extract fibers and synthetic materials. It was also true of steel-making, where the scarcity of iron or steel had to be made good by developing special metals and iron alloys. Also in more conventional sectors such as carmaking, Italian industry was distinguished by its search for compact, lightweight models, as shown by the appearance of the first cheap cars on the roads in the years leading up to World War II.

These were byways compared to the great highways of major European industry, but they allowed Italian businesses to grow and get a foothold in areas of absolute excellence. In this context Italy gained particular success in cutting-edge sectors such as aviation and telecommunications, which almost managed to overcome the general backwardness of Italian industry. Companies such as Caproni, Aermacchi, Savoia Marchetti, Piaggio, and Fiat churned out advanced aircraft often capable of beating foreign competitors and setting numerous world records. For their part, the inventions of Meucci and, above all, Marconi showed more than just the technical inventiveness of Italians. Though based on complex scientific phenomena, radio is easy to use, relatively cheap, "warm" in its relationship to the listener, and basically capable of making modern technology a part of everyday life. It was no accident that the radio was the first domestic appliance to become common in Italian homes, a first chink that was to open into a modern consumer market and encour-

Drawings and sketches
by Makio Hasuike
for Ariston Margherita,
1983.

aged the mass-production of technologically advanced products.
Alien to the spirit and space of the traditional middle-class living room, the technological origins of the radio were first camouflaged as furniture and masked in small cabinets, hidden by panels of inlaid wood and damasked fabric. The radio's gradual achievement of an independent formal identity was the result of unblinkered reflection on its design as a specific technical object, with the roots of Italian industrial design embedded in its development. The first designer to subject the radio to rigorously scientific study was Franco Albini, with his inflexibly analytical approach.
Fascinated with aviation, Albini knew just how much turns on a design's ability to separate out each component and optimize its performance and how much this contributes to the natural elegance of the technical elements. In a prototype produced in 1938 for the Wohnbedarf design competition in Zurich, Albini lucidly separated the block of circuits from the loudspeaker cone, and then recomposed them between two transparent slabs of crystal. Provocative and illuminating, the prototype of 1938 reveals a working method that had all the force and abstractness of a theoretical axiom. Albini's ideas were

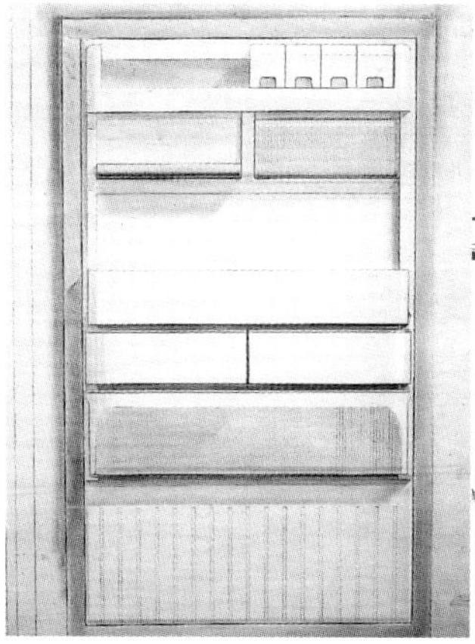

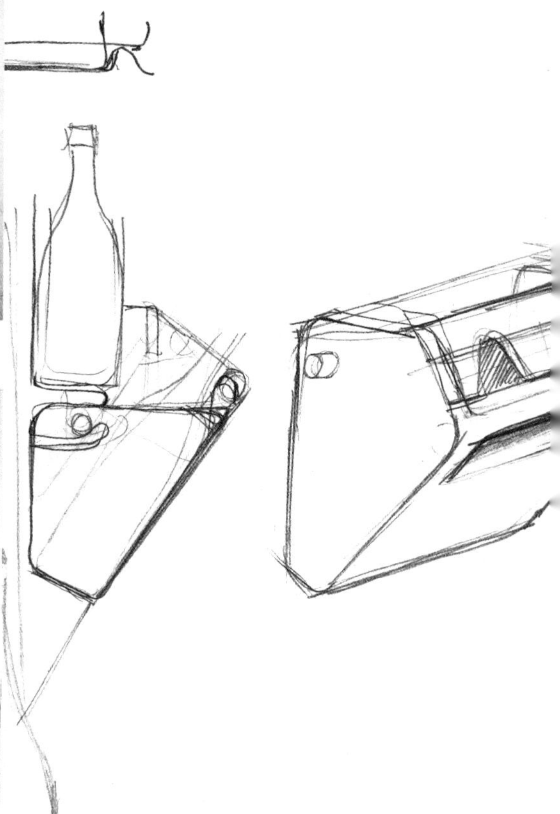

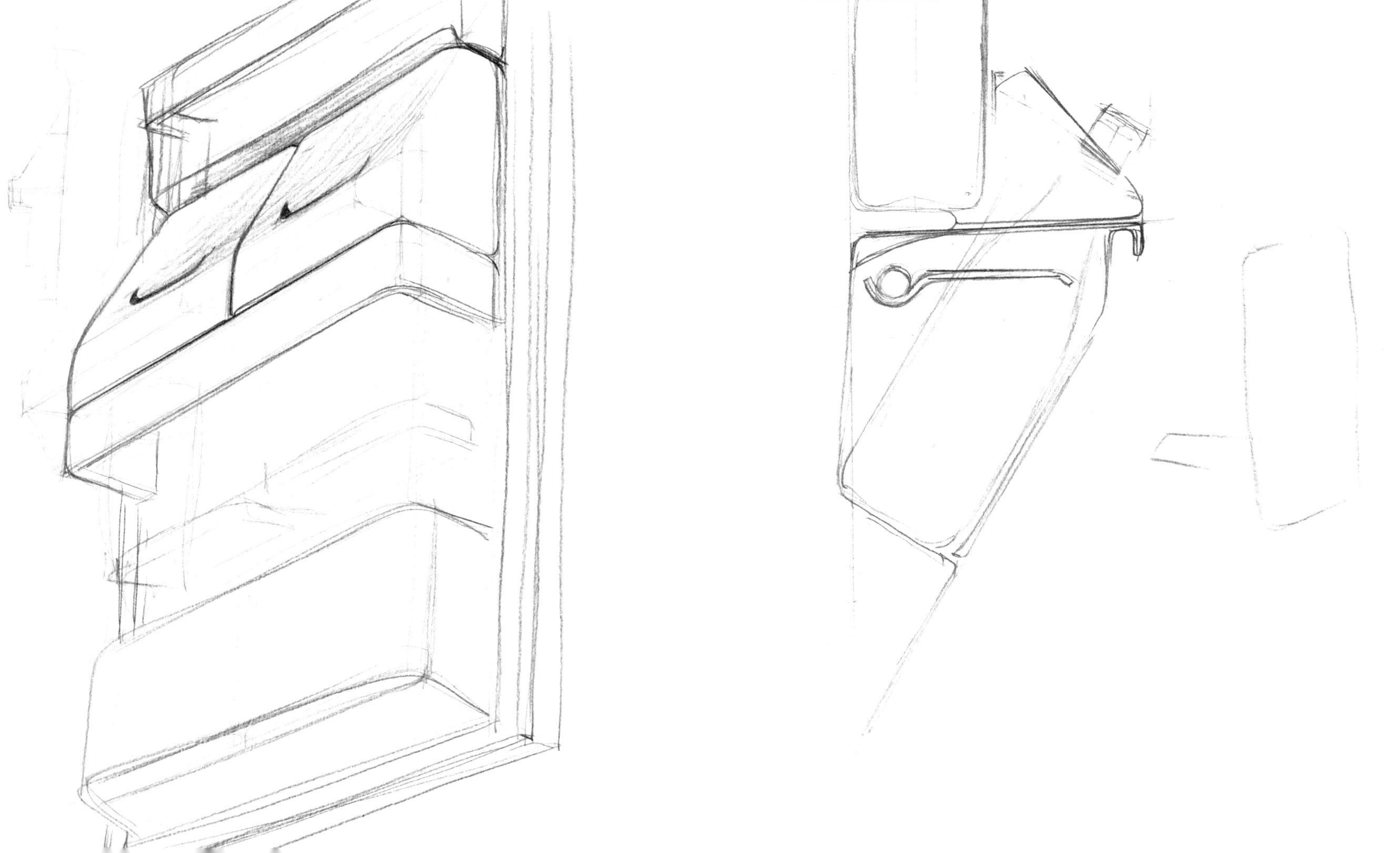

Drawings and sketches
by Makio Hasuike
for Ariston Quattrostagioni,
1987.

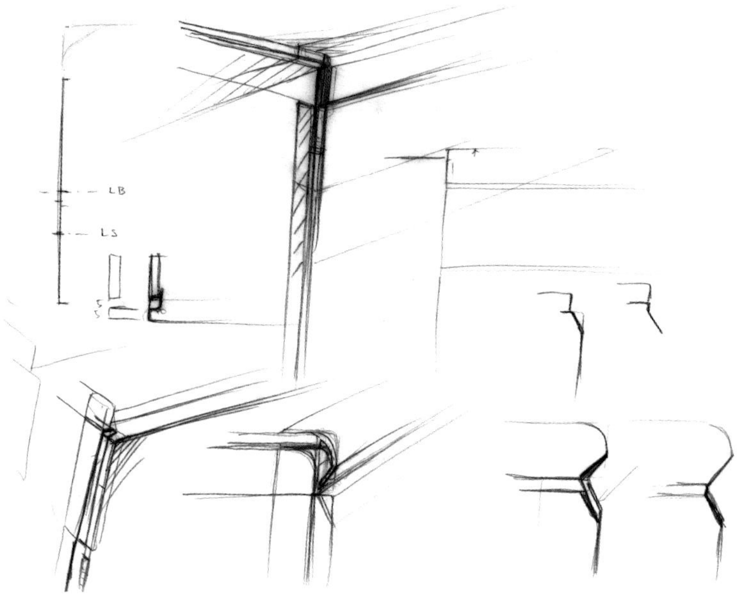

Sketches by Makio Hasuike
for Ariston Sette Cuochi.

taken up by a group of young Milanese designers—the Castiglioni brothers and Luigi Caccia-Dominioni—who just two years later designed the famous Phonola 5-valve radio. Rather than limit themselves separating the components of the radio, the young designers exploited the ductility of plastic to design a cover that displaced in space the various elements according to a precise functional program. At the front they placed a true instrument panel with the controls and the frequency panel; behind, but raised so as to project the sound towards the listener, was the loudspeaker cone; while all the electric circuits were embedded in the base.
Though embryonic, the little Phonola 5-valve radio marked the definitive emancipation of the machine from furniture, suggesting a precise methodological horizon for Italian industrial design. First the project was developed by starting from an analysis of the technical components that also redefined their spatial relationships. Second the design investigated new ideas for use, deliberately relating the machine to people and its setting. Finally, in a country without raw materials and faced with shortages in a strictly self-sufficient economy, the design sought to contain costs, reduce the weight and scale down the volumes.
But it was necessary to wait till after the II World War to see these principles applied to the manufacture of other products.
The conquest of freedom and democracy left Italy with a big bill to pay. Wartime destruction had reduced the country's stock of housing from 31 million rooms to 25 million; 8000 out of the 22,000 kilometers of the country's road network were destroyed, with the loss of all its main bridges and viaducts. Similar damage had hit the railroads, electricity and telephone lines, while industry saw production capacity cut by 40%. The victorious Allies also imposed stringent conditions: the abandonment by the Italian aeronautical industry of the growing sector of the large transatlantic carriers, the limitation of the automobile industry to the small and medium cylinder sizes, the

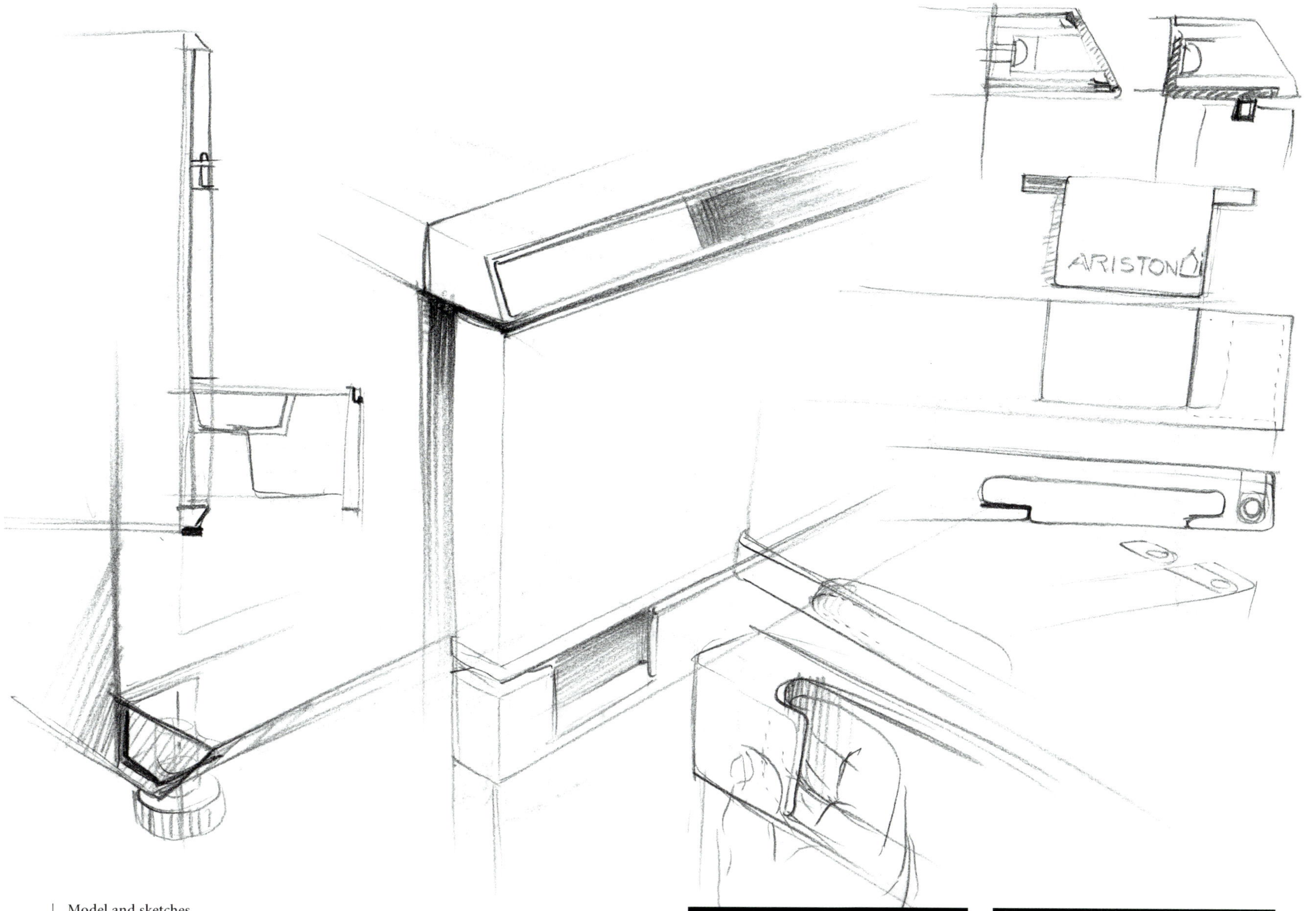

Model and sketches
by Makio Hasuike
for Ariston Supermarket,
1986.

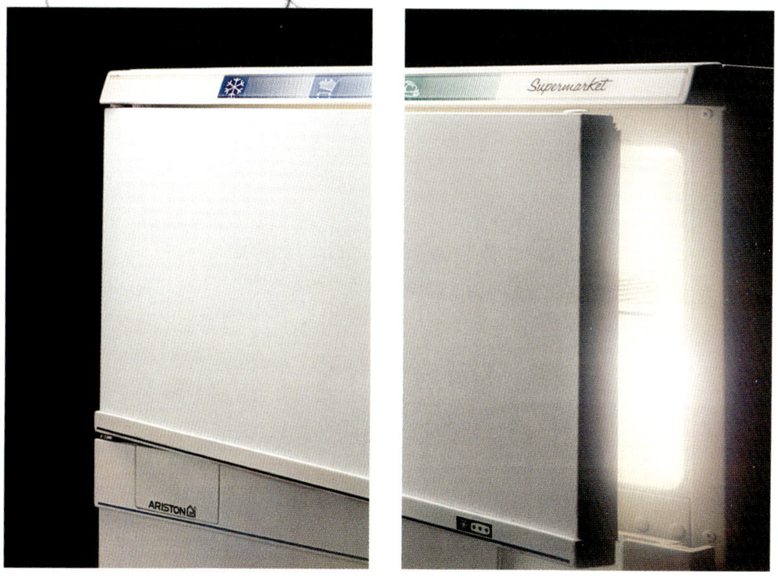

Sketches by Makio Hasuike
for Ariston Mulinello, 1988.

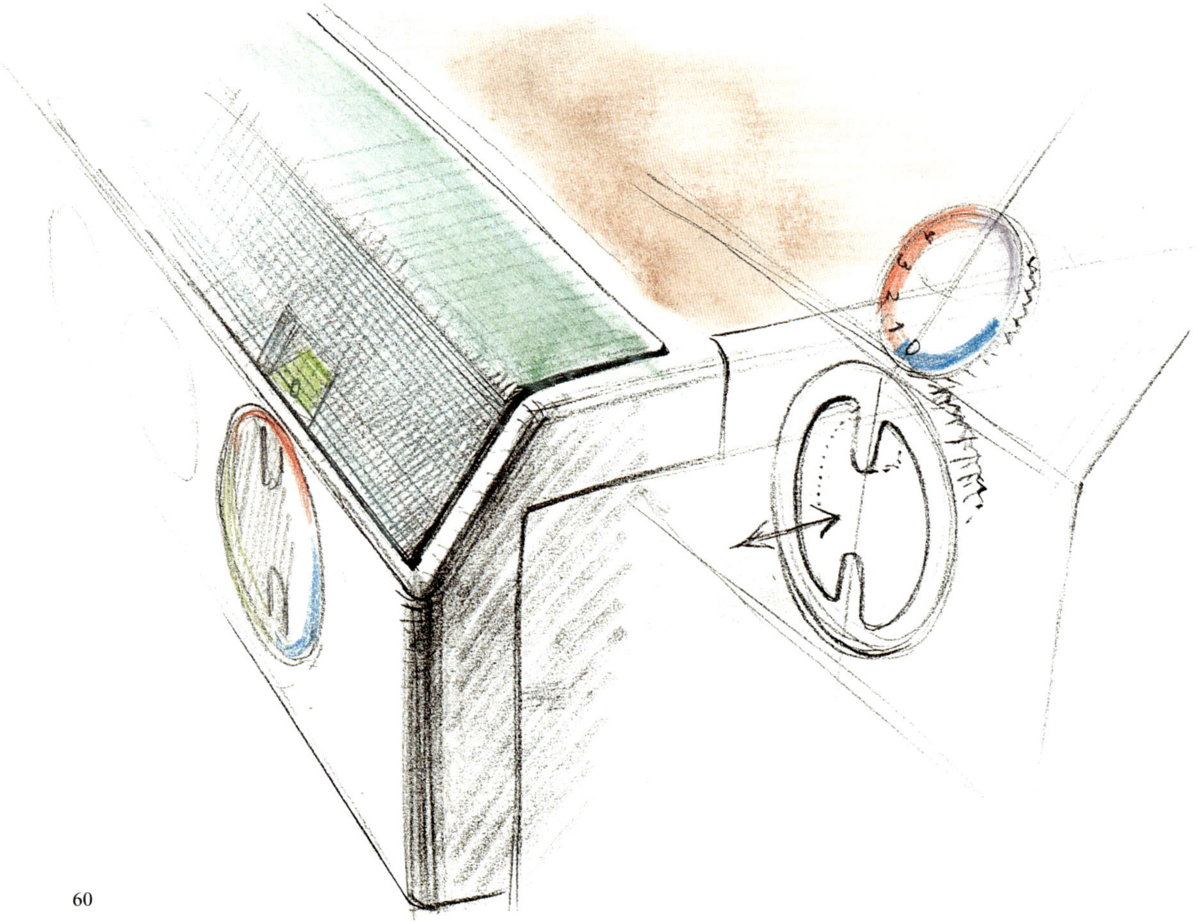

Model and sketches
by Makio Hasuike
for Ariston Mulinello, 1988.

drastic reorganization of military industries. At the same time, however, the launch of imposing programs of support to production gave a rapid start to a powerful process of reconstruction that sent the country along the path to modernization, which it had always deferred.
The restructuring of the old industrial complex and the dismantling of wartime industry freed skilled labor, capable of handling complex and advanced technologies and all at once available for new forms of production. At Piaggio, which had to drop production of aircraft and helicopters, Corradino d'Ascanio devoted his aeronautical skills to designing the Vespa (1946); at Innocenti, Pierluigi Torre and Ferdinando Innocenti adapted steelmaking technologies to the conception of the monobeam frame of the Lambrettta (1947): different technologies but both orientated towards a single and original model of individual mobility.
But it was in small-scale industry that Italy found the right dimension to transfer not only technologies but, above all,

Models and sketches
by Makio Hasuike
for Ariston Gilda, 1990.

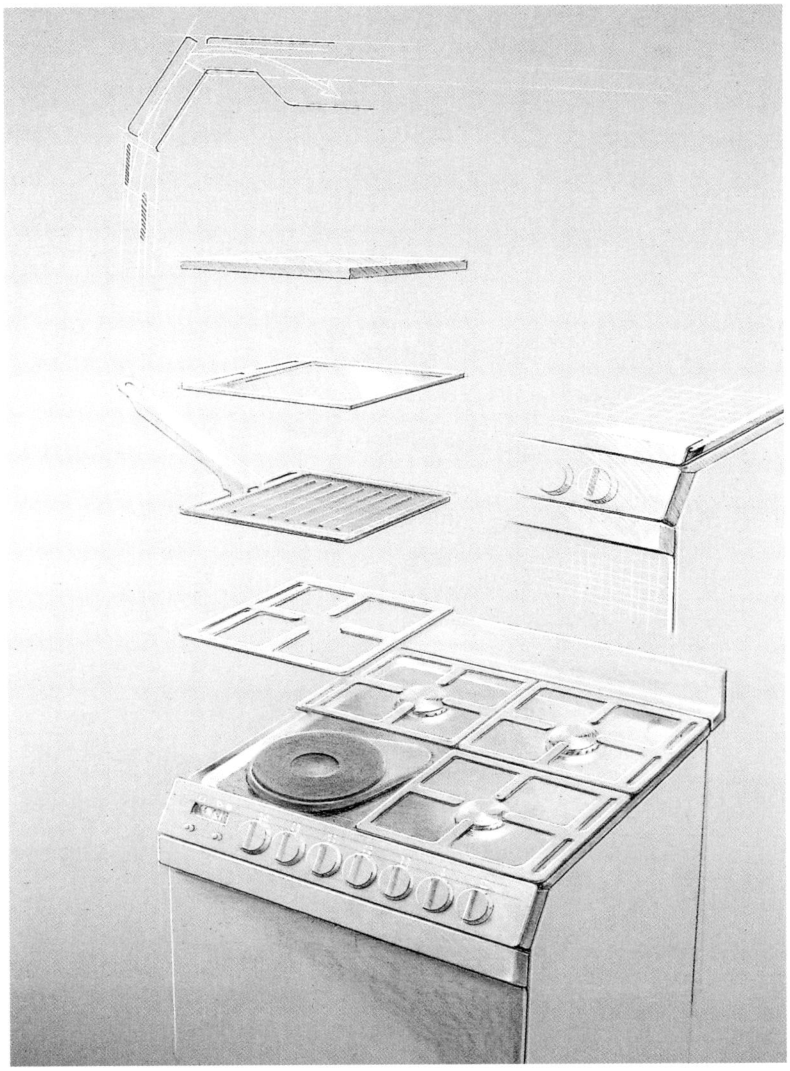

the enthusiasm of young, often first-generation, entrepreneurs and equally young designers, almost always architects with a Functionalist background. Here the embryonic researches of the prewar years found a field for experiment, enabling design to systematically anticipate market demand, surpass the expectations of consumers, and introduce technical and formal ideas that preceded the evolution of collective behaviors. It is in this context that the Italian system of design crystallized, showing how simply applying a rational approach made it possible to go beyond purely practical uses and introduce high levels of beauty, hope, technology and optimism.
Rapidly evolving social models, driven by the spread of industry and significant urban growth, set domestic design at the center of a radical process of modernization. Young architects exploited American finance to reconstruct the housing stock and started a profound revival of the very idea of the Italian home, which under the Fascist regime had remained tied to substantially traditional models. The quickening rhythms of city life, the altered composition of families, the new social figure of women with jobs, were the prerequisites for redefining the functional theories of design. Inevitably, the kitchen, with its concentration of utilities, equipment and symbolic values—an authentic laboratory of domestic life—soon formed the core of this transformation, priming the study of new models of living. As if mediating between the memories of the peasant kitchen, the aspiration to the middle-class living room and the Spartan rationality of modernist kitchens, there emerged the dining room/kitchenette scheme, a precarious compromise between functionality and decorum, between spaces for work and relationships, which still reflected the aesthetic prejudice against domestic technology. This distrust was eventually overwhelmed by the forceful growth in the sector of white consumer goods, led by the washing machine, an authentic object of desire for all the families in the Italian peninsula.
The Italian delay in this sector spurred the drive for improved designs. There was no single dominant district and no sole and uncontested leader in the sector. As early as 1953 Riccardo Bertolino in Turin developed a system of opposing springs that eliminated the need to use heavy ballast to control the thrust of the centrifuge.
But it took combined research to produce a progressive optimization of all the components of the washing machine,

Drawings by Makio Hasuike for Ariston Gilda, 1990.

On pages 64-65
Model and drawings by Makio Hasuike for Ariston Margherita Dialogic, 1996.

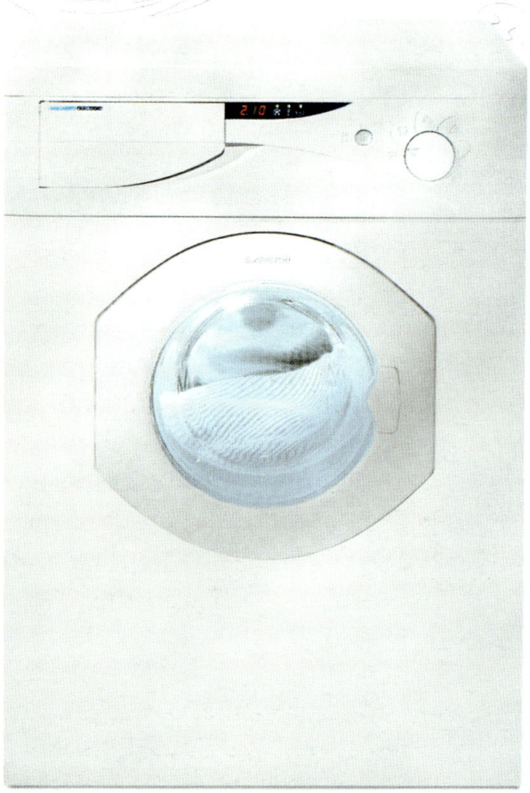

with an appreciable saving of materials and a drastic reduction in production costs. By the early sixties this process had produced a standard model of washing machine with codified measurements (60 × 60 × 85 cm), set on a base 10 cm high, with a porthole set in the middle of the front and surmounted by a functional control panel. Elegant, stylish, the washing machine was now ready to take its rightful place in the spaces of the home, without being hidden away in dark nooks and corners. Technically up-to-date; fitting naturally into the kitchen; economic in the optimization of its materials and volumes: a perfect prototype of the Italian approach to industrial design.

Aristide Merloni made his first approach to manufacturing home appliances in 1958, when the sector first revealed real scope for commercial development.

In 1958 Merloni was not an adventurous young entrepreneur: he was a captain of industry with the responsibility for a bank on his shoulders, solid political and social

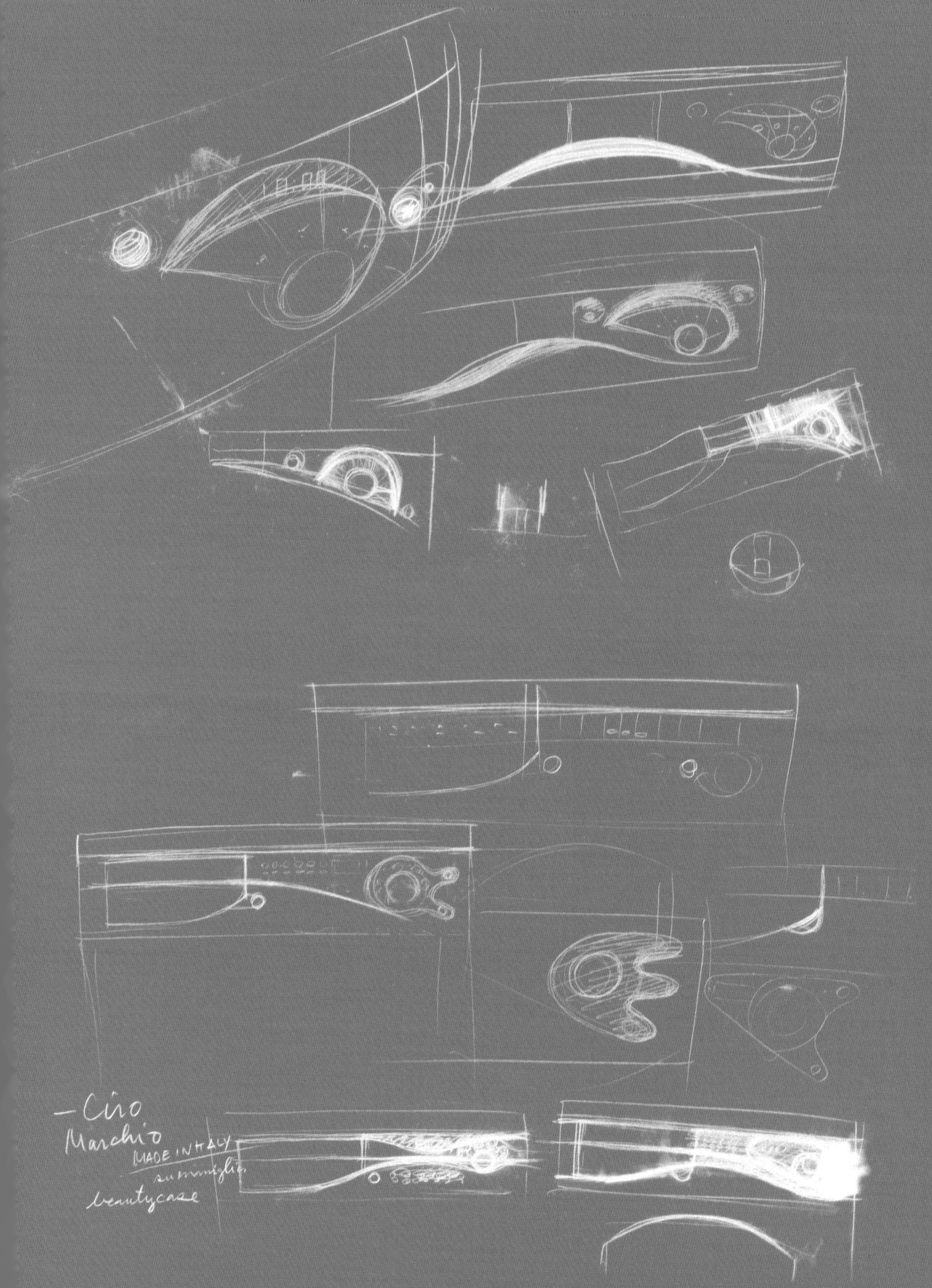
—Ciro
Marchio
MADE IN ITALY
beautycase

ON/OFF
ON/OFF
FRIDGE
RESET
ALARM
OFF
OFF
ECO
min
MODE
+
ADJUST
SELECT
–

engagement as mayor of Fabriano and a senator of the Italian Republic. Against the background of the transformations that gradually changed the face of the country in the course of the century, Aristide Merloni had already played his part as an experimenter and pioneer of Italian business. A prototype of modern man, Merloni succeeded in opening up new paths, winning his challenges and creating original business models.

Aristide Merloni was born into a family of farmers at Albacina, in the Marche, on 24 October 1897. His parents wanted a different future for him, off the land, and sent him to study as an industrial expert. When he was called up as a military engineer in World War I, Aristide was transferred to northern Italy, to what became in later years one of the apexes of the Italian industrial triangle. He found work at Pinerolo, near Turin, in the Buroni firm, which made weighbridges and scales. He had a successful career, rapidly being promoted to general manager. Instead of resting on his laurels, Merloni raised his sights and decided, in 1930, to return to the Marche and there begin his own adventure in business. They were difficult years but Merloni's tenacity, his connections, his skill at handling the human side of relations with his employees, directly and sympathetically, eventually rewarded his commitment. The firm produced scales and weighing machines, exploiting the experience acquired, and Buroni was the first of its customers. Weighing machines, in the service of industry and business, public offices and railroads, in an Italy that was searching for its road towards the future.

The outbreak of war slowed growth and the timid conversion to wartime production failed to compensate for the bombing of the factory. War's end in 1945 meant

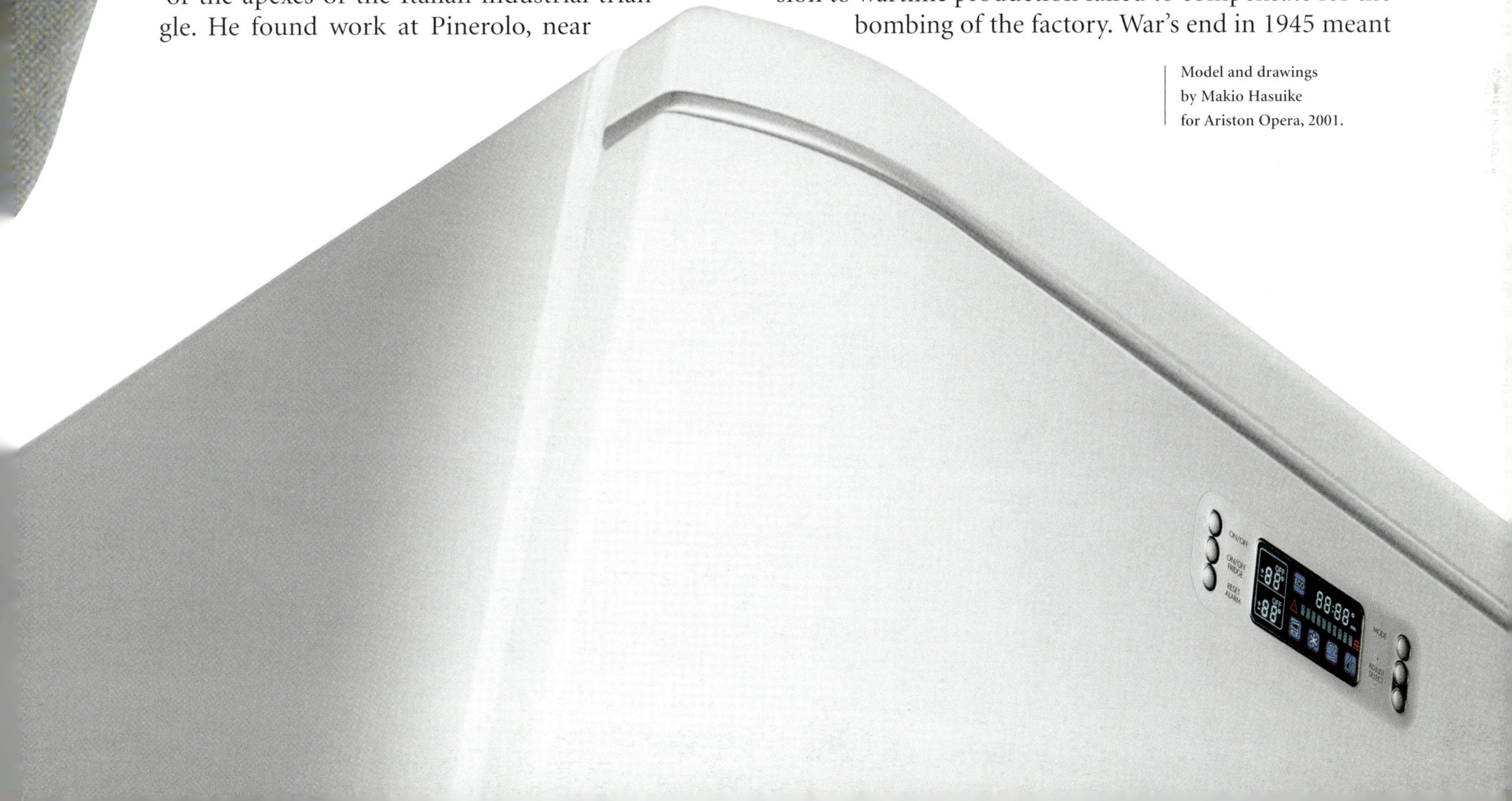

Model and drawings
by Makio Hasuike
for Ariston Opera, 2001.

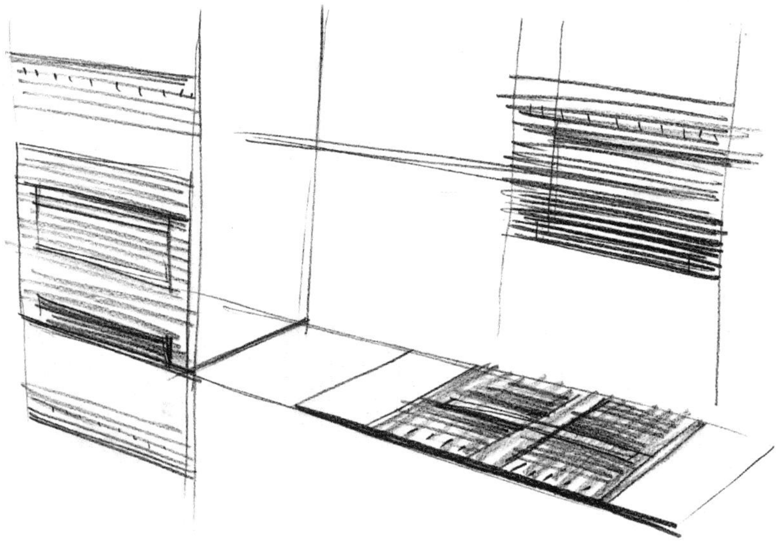

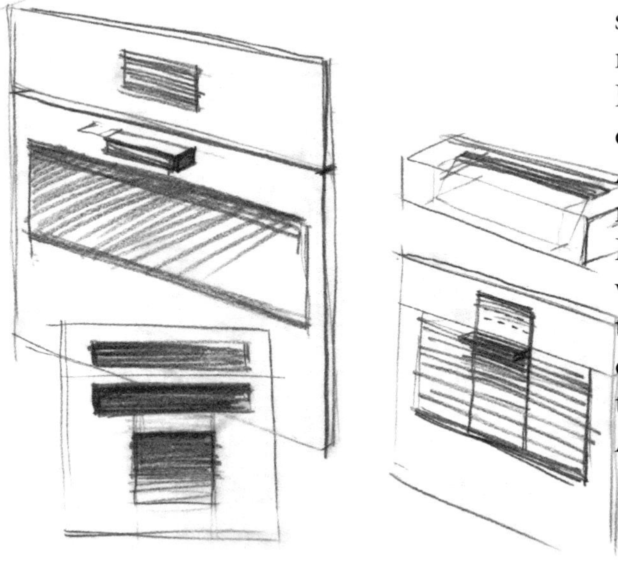

a new beginning, but reconstruction fuelled orders: Italy was getting under way again. In 1950 Merloni Industries were the leading Italian producers of weighing machines and enjoyed sound financial health. When his eldest son Francesco gained a degree in engineering, Aristide decided to diversify production and opened a new product line: liquid-gas cylinders. The choice faithfully reflected the Italy of those years: liquid gas, pure energy for the country's growing industrial muscle but on a small scale suited to the limited needs of tradesmen, small businesses and new domestic users, a privileged field for more modern individual behavior. After some initial difficulties production took off, revealing the clear insight behind the idea.
Strengthened by the results reached, Aristide Merloni insisted on diversification, convinced this was the way to multiply turnover and jobs. Seeing the rapid developments in society and living habits, Merloni gambled on the home and its modernization, flanking gas cylinders with liquid-gas stoves, cookers and boilers. Splitting the company up into independent businesses solved different problems. First of all it enabled Aristide to allow each of his two sons to control a part of the group. Then specialized production made product development and improvement more efficient; the relatively small size of each sector prevented the inefficiency found in overgrown firms and reduced business risks; the small production facilities were easier to fit into the socio-economic texture of the Marche, reinforcing the virtuous circle between business, countryside and workers that had always been at the heart of the firm. Merloni significantly used the term "metalworking sharecroppers" to describe the close relationship that the business established with its setting and workers. Organized as a "galaxy" of initiatives, the different distributions were coordinated and combined under a single brand, the Ariston logo, from the Greek word meaning "the best", registered as its trade mark in 1958.
Flexibility, innovation, research: the characteristic ingredients of Italian design were already all implicit in the Ariston model. They were applied in the technical departments of the business as well as in the farsighted Aristide Merloni Foundation, an R&D center that acted as the fly wheel for innovation and research for both the group and the region. But to establish a deeper and more complete connection between the business and modern design culture there had to be that generational handover that Aristide had wisely begun. The professional growth of the

Model and drawings
by Makio Hasuike for
Ariston Linea Experience,
2004.

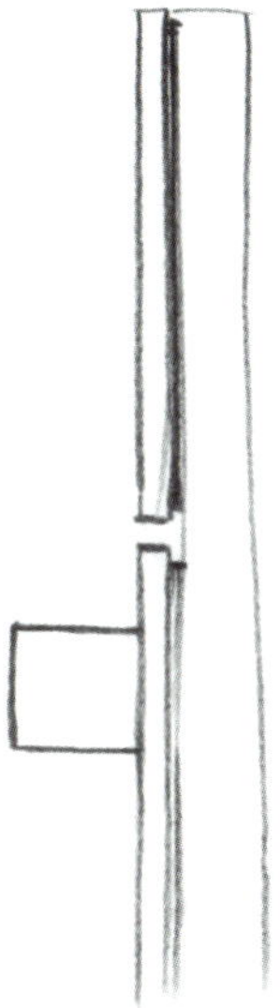

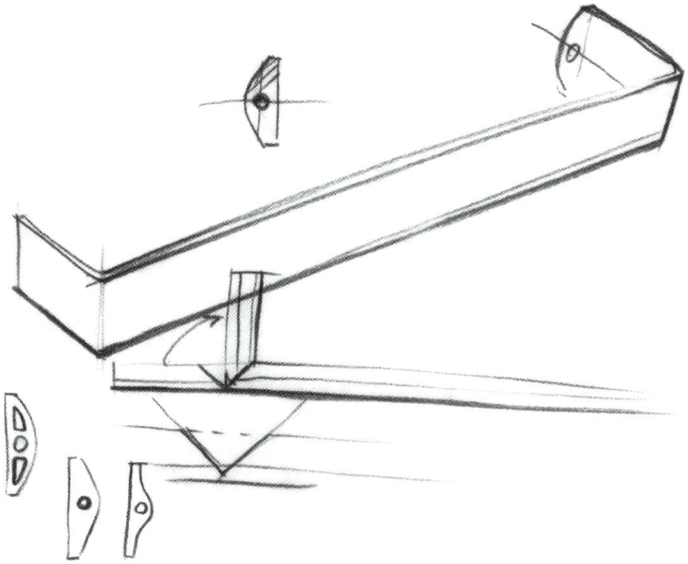

Giorgetto Giugiaro.

Model and sketches by Giorgetto Giugiaro for Indesit washing machine, 2001.

younger of his sons, Vittorio, who joined the firm in 1960 with responsibility for home appliances, laid the foundations for this new awareness.

The key moment of this route was to be the meeting between Vittorio Merloni and Makio Hasuike: "I first met Makio Hasuike in the mid-sixties… It was… his professionalism that immediately told us we were not in the presence of a creator of more or less abstract forms, which would create difficulties when they had to be adapted to the reality of production… Makio understood the technical problems behind the creation of a product… He understood that elegant design had always to get to grips with the functional nature of the concrete object… At Ariston we came to an agreement immediately at the first meeting. It proved a full and durable agreement, with the partnership continuing unbroken after more than thirty years."[1]

The basis of this partnership was necessarily very solid, but also sufficiently elastic to accommodate change and transformation. The ability to listen, to get to the bottom of the constant process of innovation, but above all curiosity and an ability to always ask new questions were the gifts that enabled Makio Hasuike to accompany the growth and changes in a large group like Ariston. Born in Tokyo in 1938, son of artist parents, Hasuike approached design through the study of Functionalist ideas in university textbooks, but he fell in love with it in the pages of Italian maga-

Model and sketches
by Giorgetto Giugiaro
for Indesit dishwasher, 2001.

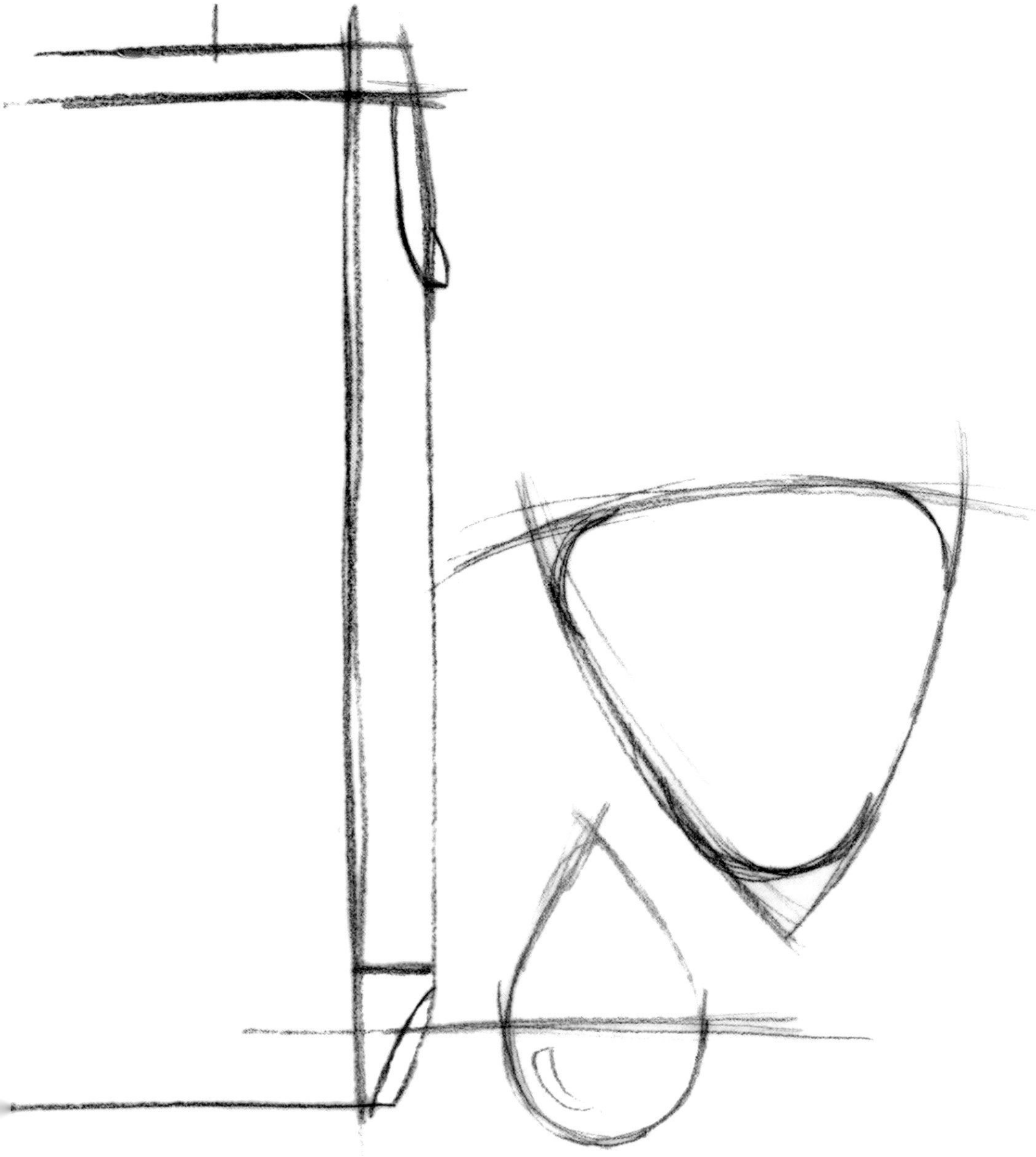

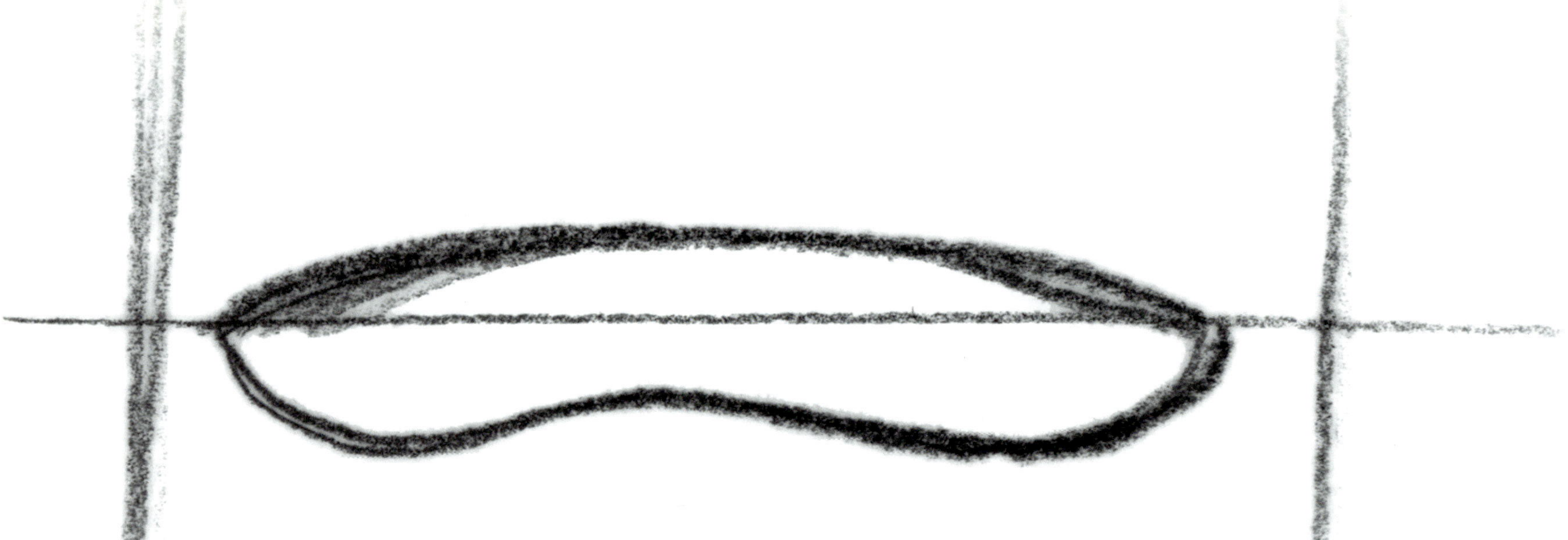

zines, where he learnt about the free experimental research being conducted at Olivetti by Nizzoli, Sottsass, Zanuso, Bellini. After a year working for Seiko designing the dials of watches and chronometers for the Tokyo Olympics in 1964, Hasuike decided to visit Italy. Paradoxically, once in Milan, his career began in the firm of Rodolfo Bonetto, standard-bearer of the Ulmian Rationalist tradition. The tension between austerity and freedom became a constant in the work of Hasuike, a designer capable, like few others, not only of developing ideas but of asking questions and stimulating innovation. The 1968 was a crucial year in Hasuike's career: guided by Dr Lacanina—head of La Rinascente's buying department—he began working for Gedy, a business that specialized in the production of small bathroom accessories, and above all with Ariston.
His first contact with the firm came when he was completing a research project, devised in 1962 by the Aristide Meloni Foundation, to unify the firm's different kitchen appliances as a single unit. The commercial launch of the, simple, functional Unibloc system (1968) introduced a new image, somewhere between the starkness of the laboratory and the domestic universe, finally foreshadowing the reunification of the dining room with the kitchenette. As Hasuike recalls, it was "a plain sink set in a long steel plane which grouped together underneath the three home appliances set side by side."[2] The design was actually rather more complex and incorporated appliances and cupboards below the sink in varied compositions of different lengths, rationally integrating furniture and functional equipment in a way that superseded the traditional hierarchy between furniture and machines. This was especially important in a firm whose divisions were already tooled up to produce all kitchen components, both furniture and appliances.

Model and sketches
by Giorgetto Giugiaro
for Indesit refrigerator,
2001.

Inscribing the formal vocabulary of the appliances within the domestic scenery was Hasuike's first task. Working with Ariston's different departments, he began a general formal and functional review of production. Emblematic of this were two products: the U10 electric water-heater (1969), elegantly rational, conceived to be set in full view next to the sink, and the Aristella dishwasher (1979), the first to dominate the Italian market. The dishwasher was faced with colored laminate panels and fitted in nicely with the kitchen furnishings. Hasuike, however, continued to explore ways of unifying the various elements of the design kitchen, at the same time superseding the rising dualism between free-standing and inbuilt elements. The general adoption of the continuous work surface incorporating the sink and cooker hob, a technical surface but enhanced by a thickness of at least 7 or 8 centimeters, had by this time introduced a sort of elegant formal horizon in the kitchen: beneath it were placed the technical elements and the machinery, camouflaged behind facing panels, and above it rose the nobler furnishings and wall-hung cupboards rising towards the ceiling.

Sketch by
Giorgetto Giugiaro
for Indesit cooker, 2004.

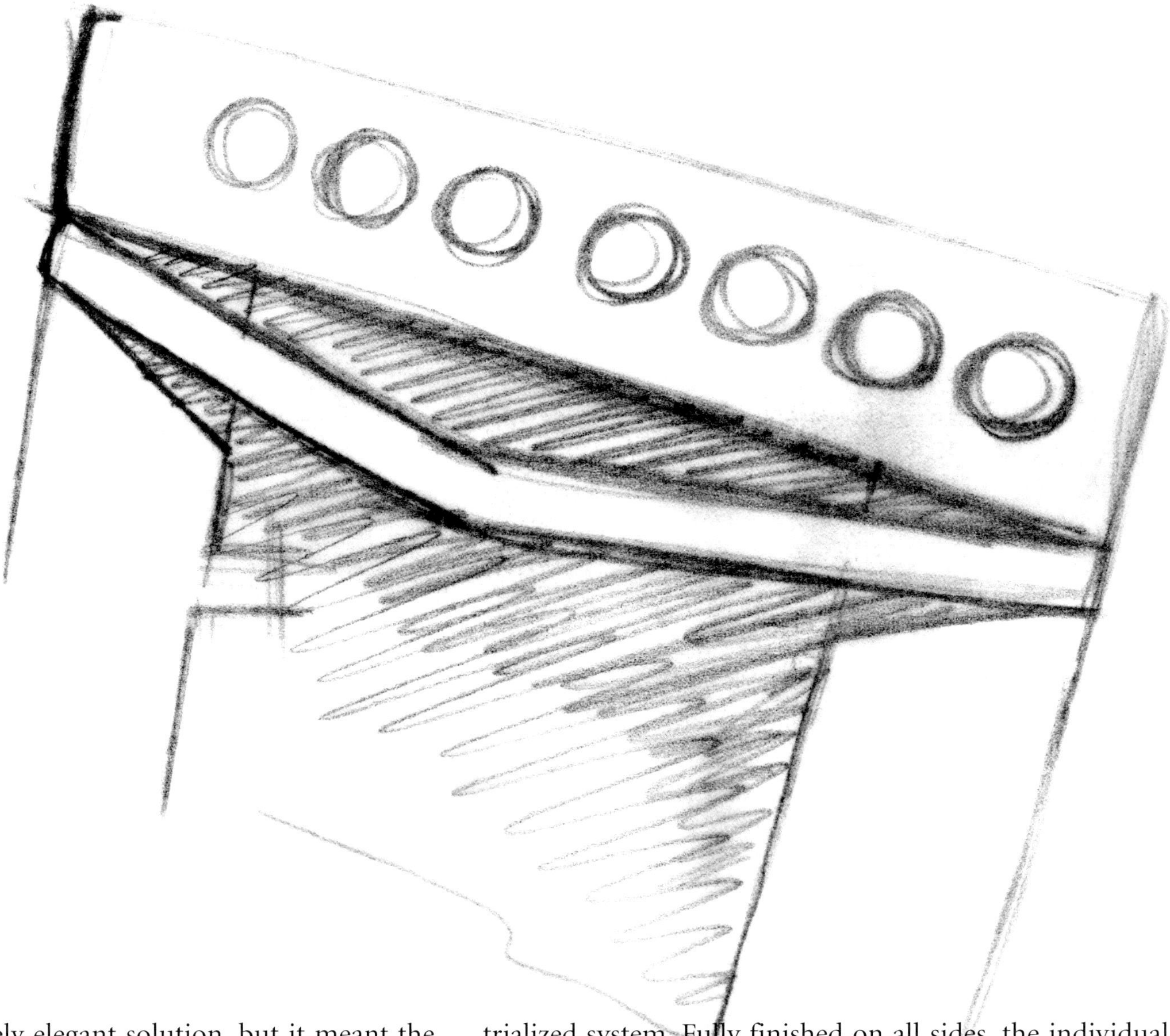

This was an extremely elegant solution, but it meant the whole kitchen had to be bought as a single block, complete with all appliances. Ariston and Hasuike judged it risky to downgrade the appliances to appendages of the furniture and stake everything on a system that called for a substantial initial outlay. "I thought it would be possible to spread out the investment by marketing a series of free-standing… but perfectly coordinated products. The plan naturally included both appliances and an extensive range of wall-hung and floor-mounted cupboards."[3]
After three years of development, this project was presented in 1977 under the acronym OSA (Open System Ariston): with standardized dimensions (a single module measuring 60 × 60 × 82 cm), coordinated components and structures, a complete color code, introduction of parts in expanded resins and aluminum to form a complete industrialized system. Fully finished on all sides, the individual elements could be combined quite freely, in part because all the utilities were set at the back. The design won the Compasso d'Oro award in 1979, but never had the hoped-for commercial success. All the same, the firm distilled the experience accumulated in creating the project into two parallel divisions, one for built-in appliances and the other for free-standing units. "In practice the two kinds remained quite distinct, covering two different slots of the market. The free-standing system became more clearly basic… while the fitted-kitchen collection was more securely lodged at the top of the range."[4] Hasuike fostered the firm's commercial strategy, which avoided small production runs of exclusive products as well as the cheapest, low-quality appliances, but he also explored alternative ways to develop the image of the kitchen. Though designed

Model and sketches
by Giorgetto Giugiaro
for Indesit cooker, 2004.

as a built-in system, the D1 fitted kitchen of 1981 departed from the "monolithic austerity of the integrated system by creating a light and colorful kitchen landscape, with airy shelving and suspended nets". The following year he designed the Modulo program with coordinated units that established a new standard of quality for systems furniture and free-standing appliances.
The eighties marked a turning point in many ways: Italy permanently joined the cluster of affluent societies and new patterns of consumption emerged. The opening pages of magazines celebrated "intelligent" vacations; hi-fi systems, color TVs and video recorders became common. In bathrooms there was a fashion for hydromassage tubs, the cutting edge of a revolution that was to transform the image of the most secret room in the house in the years ahead.

Sketches by Giorgetto Giugiaro for Indesit washing machine, 2003.

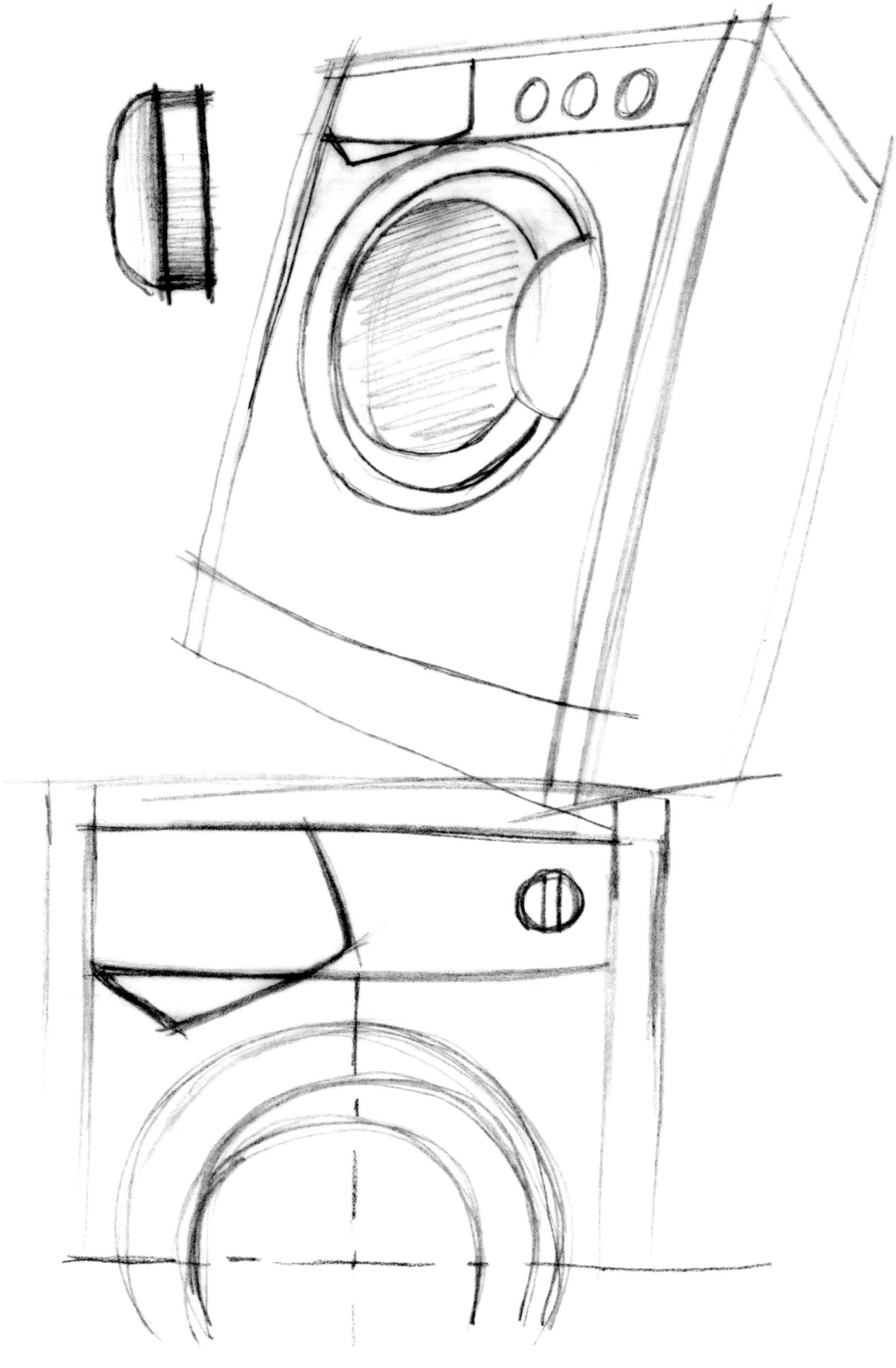

In the kitchen, affluence became abundance, transforming the fridge into a generous-sized distributor that the freezer, a new entry on the market, seemed likely to render eternal. Hasuike worked in successive models, redesigning both the inside compartments and the structure of the varied temperature cells, as if imitating in vertical succession of hatches and drawers the pattern of an eighteenth-century *sécretaire*. But nothing seemed to really enhance the monotonous uniformity of the forms of electric appliances. In tackling the problem of their standardized shapes, Hasuike intervened with extreme delicacy, without excess and without letting himself be tempted by the facile sirens of postmodernism. His design of the Margherita washing machine of 1983 is a tangible example of the equilibrium with which Hasuike worked on what was already an archetypal and established object. He simply concentrated on the design of the porthole, almost an anthropomorphic eyelid, and on the clear graphics of the various controls ranged in a strictly geometrical band. Twenty years later, the design is untouched by the passing of time.

The eighties also marked a turning point in the firm's strategy. In 1987 it was quoted on the Milan stock exchange and started a campaign of important acquisitions that has continued in recent years. Also in 1987 it took over Indesit, an established Italian brand which had been struggling for some time and eventually gave its same, Indesit Company, to the group; two years later it was the turn of the French brand Scholtès, preluding a significant expansion on foreign markets with the recent acquisitions of the Russian brand Stinol and the British Hotpoint in 2002.
This aggressive commercial strategy required the design team to support its multi-brand policy with endless vari-

Model and sketch
by Giorgetto Giugiaro
for Indesit dishwasher, 2003.

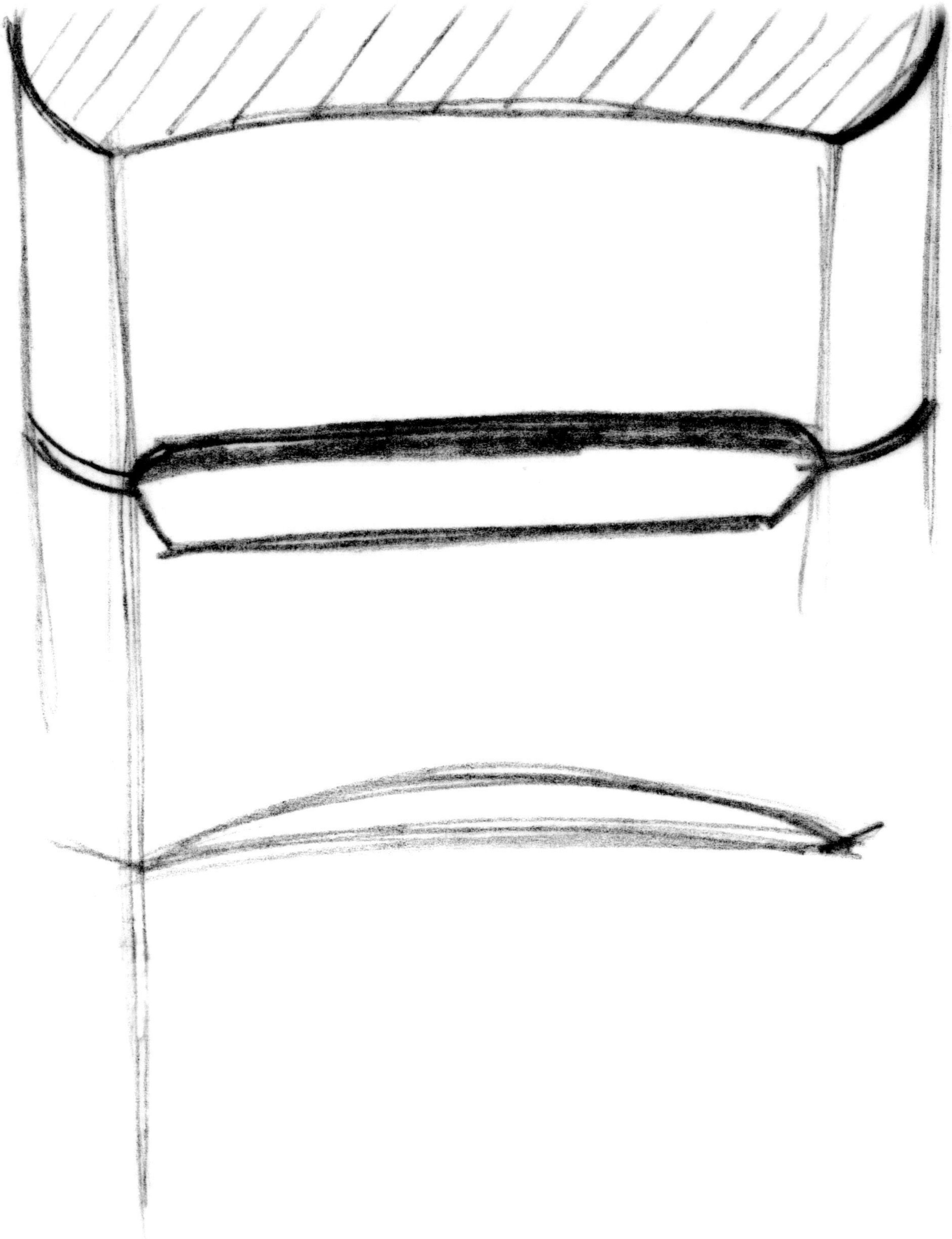

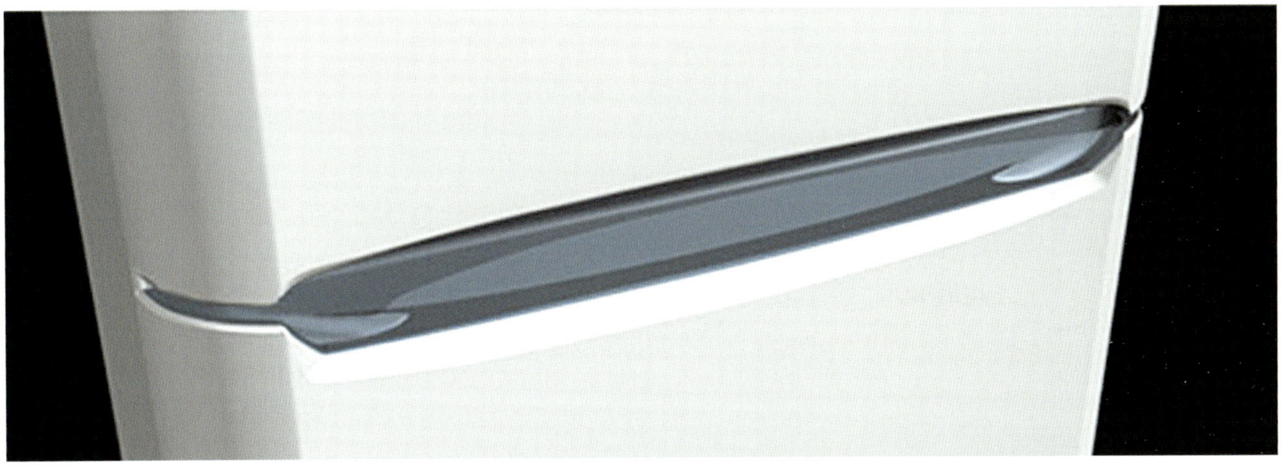

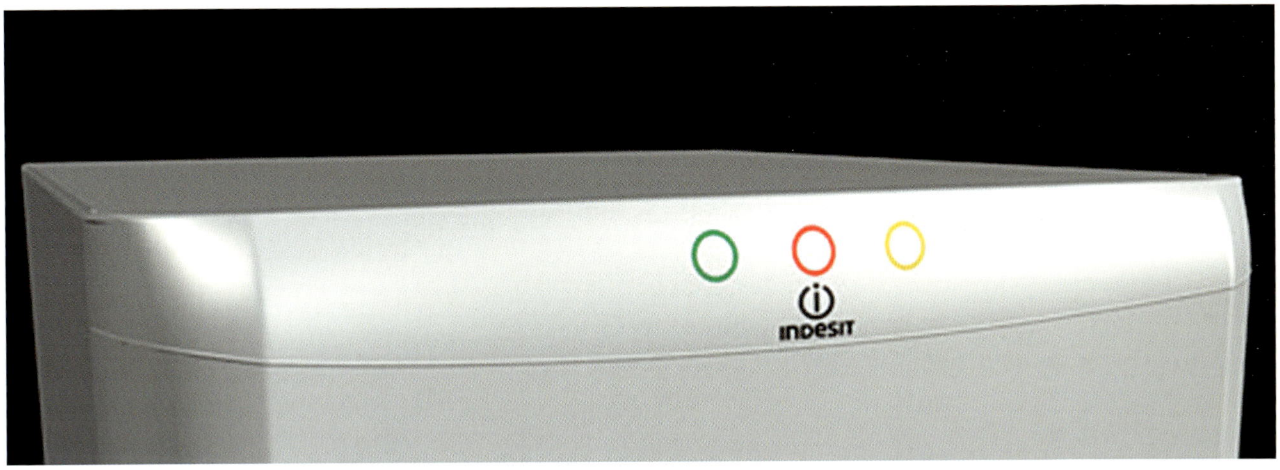

Model and sketches
by Giorgetto Giugiaro
for Indesit refrigerator,
2003.

ations. It boosted research, particularly into the new electronic technologies now entering domestic appliances, with the spread of personal computers. The objective was to study the scope for enhancing the performance of appliances by no longer redesigning their functions alone but giving them the ability to accumulate and manage their own "intelligence and experience", and so offer consumers more highly developed performance. In truth, even in the early seventies Vittorio Merloni had launched an Electronic Applications Center to study integrating the mechanical core of the machines with a digital control interface. But the costs of electronic components and the lack of connection between the two systems had led to its abandonment and the Center was closed. It was only at the end of the following decade that the progress made by electronics in the automobile sector and the appearance of innovative Japanese appliances suggested taking a second, more careful look at the new technologies and setting up a new Research Laboratory, the heart of electronic research at Indesit Company. Alongside actual research, it also began to foster a true and original electronic culture within the firm. This meant that the new technologies would not simply replace the conventional electromechanical functioning of appliances but create real innovation in their performance and functions.

The starting point for research was the analysis of a special theory of logic developed in 1965 by Professor Lofti Zadeh and called "fuzzy logic", as an alternative to the Cartesian binary code (true-false, on-off, empty-full). This made it capable of coping with intermediate states that separate extremes in the field. Six years were to pass between the first steps of the Research Laboratory (1990) to the marketing of a digital appliance. They were studded by a lot of studies, advanced courses in the United States, various experiments and numerous patents. The key to the innovations developed by Ariston's laboratories was an original system of integrating cutting-edge electronic and mechanical components with low-cost sensors, which had always been present in electrical appliances. The responses of the sensors feed information to the electronic components—suitably programmed with fuzzy logic—enabling them to extrapolate informations that make the machinery automatically select a whole sequence of operations and controls. A first experimental application was developed in the workings of the Quattrostagioni refrigerator (1993), which could adapt the temperature of the internal compartments to a series of preset parameters, while also receiving orders from a personal computer. It was the first appliance that could be linked up with a network but above all it provided a useful opportunity to get all sectors of the group working together in the same field of innovation. The results led to the project for a digital washing machine. Launched in the summer of 1995, the Margherita Dialogic model could process a large quantity of information furnished by three simple electro-mechanical sensors supplemented by programmed electronic circuits.

The introduction of electronics was not exhausted however in digital control of the appliance to guarantee better performance and ease of use: the electronic components generated and stored information about the functioning and the habits of the user as well as transferring this information to the exterior through a line of serial communications. Heady prospects in years dominated by the myth of the new economy, when it seemed the digital revolution would generate enormous earnings by radically and swiftly transforming all our relationships with the universe of objects. Vittorio Merloni's decision in autumn 1996 to entrust the future of the firm to a new managerial team directed by Francesco Caio provided the opportunity to impress a marked acceleration on the digital development of the products.

All the potential of the research in course was explored and mobilized, foreshadowing the interconnection of the network between the different appliances and their connection with internet technology. The enthusiasm led to the creation of an independent company (Wrap—Web Ready Appliance Protocol) to incorporate the research laboratory. The research was focused on various objectives: regulating consumption to suit the load (Ariston Digital, 2001); controlling and monitoring the operation of each single appliance from a distance (Margherita2000.com, 2000); preventive maintenance suited to the real level of use of an appliance; theorizing a system of leased appliances with rates based on use (Pay-X-Use, 2002). These projects were all completed but in actual fact they proved futuristic rather than real domestic systems. The absence of a standard protocol for linking home appliances in a network, accepted and adopted by all manufacturers, as well as the high cost of wiring up the home for a dynamic, interactive system, indispensable for any possible development of domotics, are also obstacles to Wrap's latest project, a small device capable device of tailoring the consumption of each appliance to the availability of the network. Despite the difficulties in the way of connecting integrated systems of digital appliances in a network, the research carried out in the group produced important financial results, as well as burnishing its image. The development of a single electronic card, shared in 2003 by over 500 commercial models of washing machines, made possible immense economies of scale and appreciably cut the cost of the components with the highest technological content.

Undoubtedly the versatility of the digitally controlled devices required the designer to develop different brand and range identities, definitely very important within a group articulated in different brands, each with its own distinctive identity.

In the case of appliances branded Ariston, strong with an identity firmly established over the years, Hasuike adopted a strategy of continuity, in some cases—emblematic that of the Margherita washing machine—going so far as to preserve the image of the product though it was now "fully wired" with new technological contents. In general, though, his task was to evolve the project by incorporating new functions in forms compatible with the changes in domestic imagery, emphasizing the technical elements to foster a hi-tech trend or, on the contrary, dematerializing the components in a rarified minimalist horizon. These are apparently immaterial concepts yet they interact with domestic behavior, which the technology makes concrete and practical. As in the case of the SuperSilent washing machine, which reduced the noise threshold by adopting an engine that works on three-phase energy. A simple idea, but it has been made possible only by adopting an excusive device patented by the company, which converts the normal mains voltage inside the machine.

It really is SuperSilent, so it can be used at night when energy costs less.

The task of taking the Indesit brand upmarket, entrusted in 1993 to Giugiaro Design, meant identifying it with images capable of communicating a new identity and new values. Addressed to a public not necessarily identified

Drawing
by Giorgetto Giugiaro
for Indesit Built In oven,
2004.

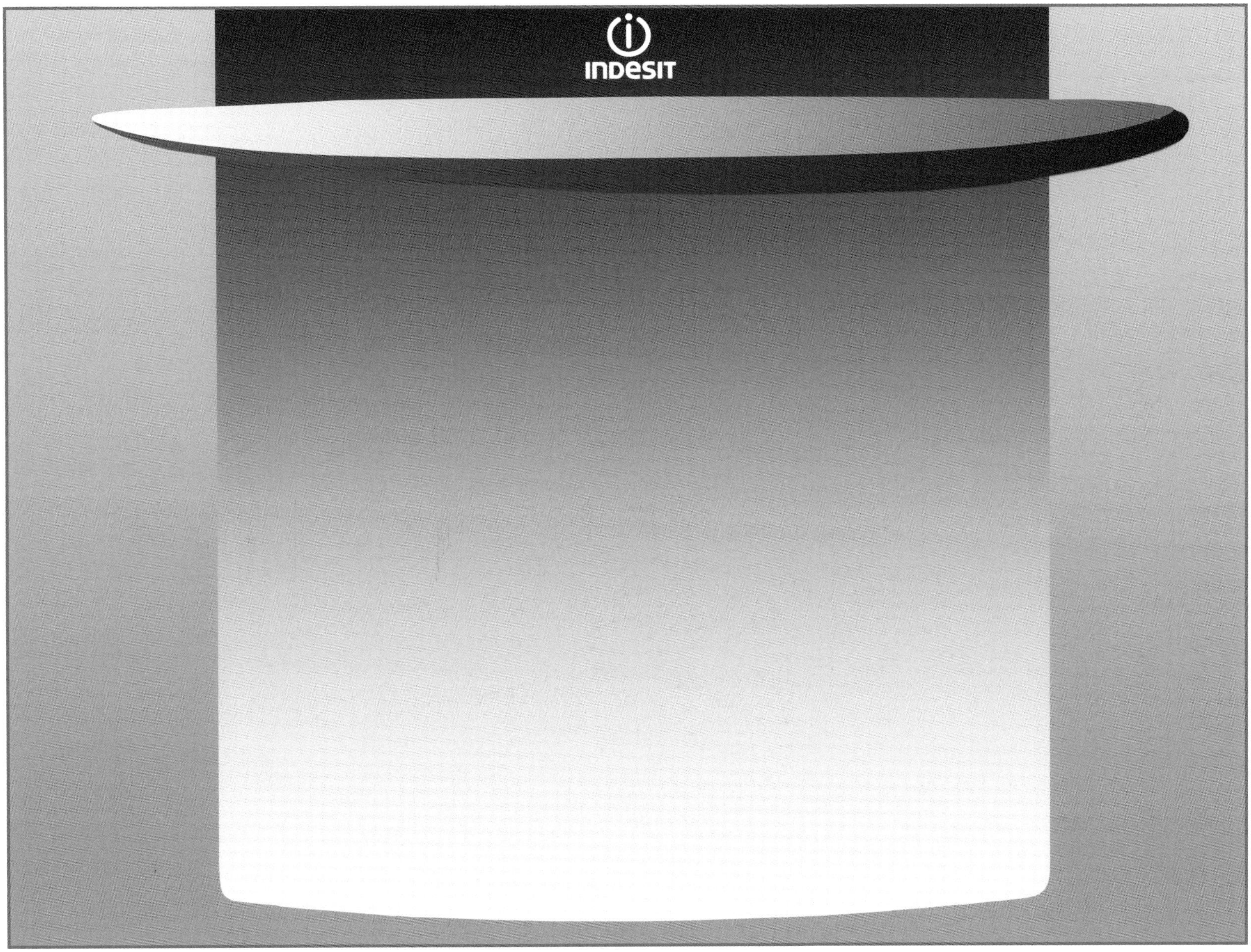

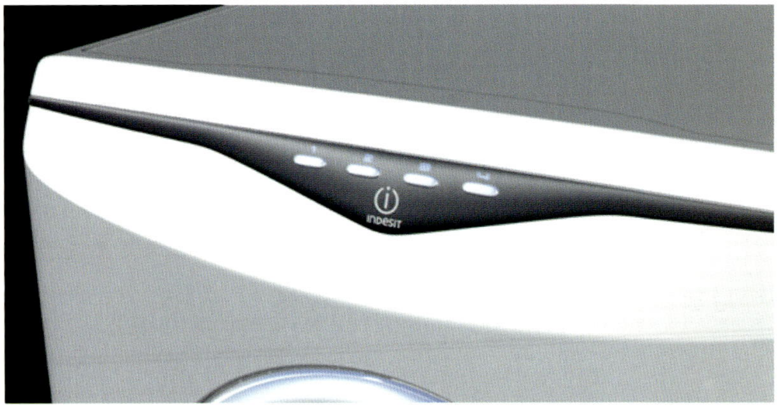

Prototypes by Giorgetto Giugiaro for Indesit Smart & Simple.

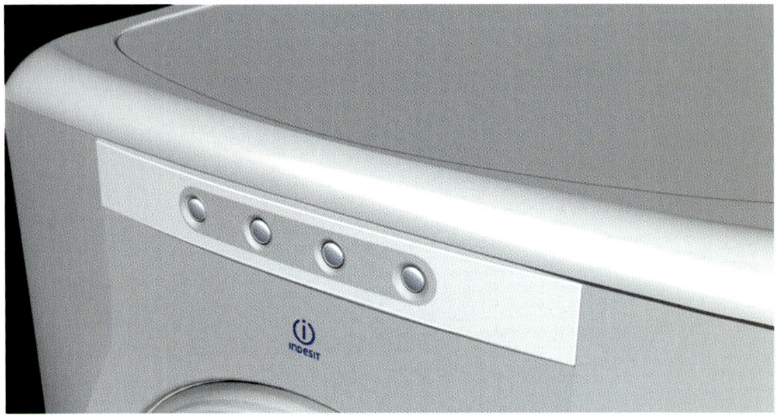

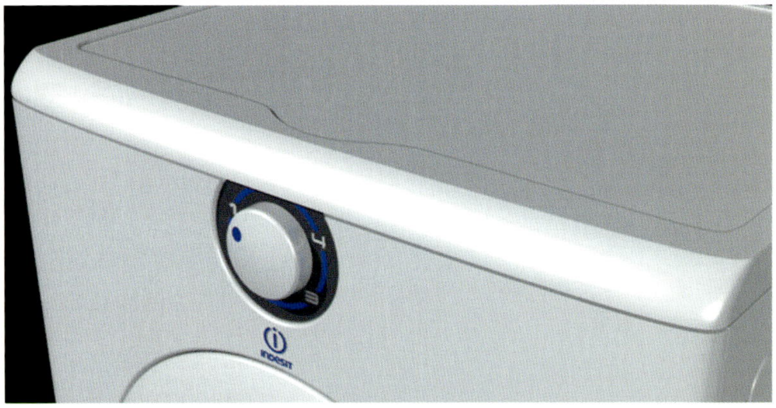

with the traditional family, the Indesit range was born digitalized and adopted solutions that privileged an immediate understanding of the highly evolved services offered by the new appliances. Having achieved maturity, as a part of everyday life, home appliances no longer need to display their leading functions so much as emphasize their new potential: not just washing, but at what time, with what load, with what options in energy-consumption, with what intensity, and how long it takes to complete a wash. Not refrigeration, but different temperature settings, multiple degrees of moisture, quick-freeze compartments, freezing portions into easy-access cubes.
All this is emphasized and made intuitive by simple machines, easily understood, specially designed to translate the range of options into intuitive and simple controls. Today, faced with an important strategic change that sees all the different brands united in the name Indesit Company, it is precisely this legacy of research, innovation, and patents that provides a solid base on which a new future will be built.

[1] Vittorio Merloni in F. Zurlo, *Makio Hasuike*, "Abitare Segesta", Milan 2003.

[2] Makio Hasuike in E. Morteo, *Ospiti in cucina*, "Arredare", 7, Editoriale Domus, Milan 1994.

[3] Makio Hasuike in *Ospiti in [...]*, art. cit.

[4] E. Morteo, *Ospiti in [...]*, art. cit.

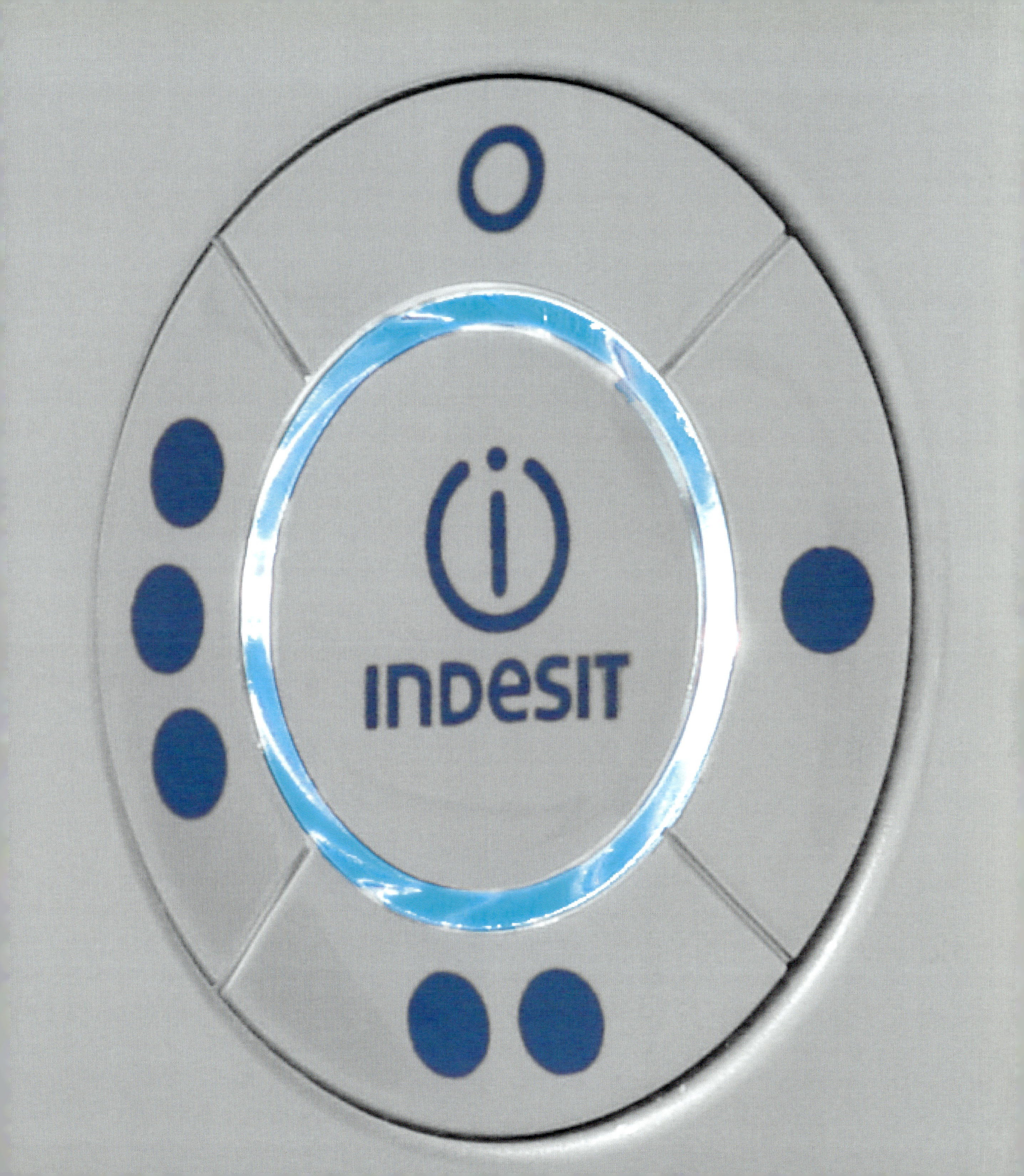
INDESIT

Sociology

INDESIT

Maria Sebregondi

"A whole world disappears and is replaced by phenomena of an entirely different kind."
Paul K. Feyerabend, *Dialogue on Method*

Our way of living is caught up in a new mutation that has all the features of a true breakthrough in the evolutionary process of the human species. *Homo sapiens* has always been a symbiont, meaning a hybrid of human and technological undergoing continuous change: the original symbionts were creatures with a low technological intensity, but in the course of the time they were gradually replaced by creatures with a more intense technology.
Domotics—the application of information and electronic technology to the home—after featuring for a decade in incredulous discourses and imaginary projections, is becoming a fact, or rather a cluster of facts of a new kind, a true network of information and connections that are changing our daily life in depth. Our individual, one-to-one relationship with the machine, understood as an isolated technological object, is gradually ending. We are seeing the end of the encounter/clash between the thinking human being and the mechanical creature, helpful or threatening, that has given rise life to innumerable tragicomic representations.Machines have learned to talk to each other, interact; they intuit, advise, remember.
A new friendly technology is approaching and will harmoniously facilitate, simplify, organize and coordinate our lives. Not longer *machines célibataires*, but conjugated, hyper-related, ready to combine and generate the great Nanny we're all dreaming about.
But where will this big Nanny come from, how will she work, what does she promise and what threaten? A set of snapshots, arranged alphabetically, seeks to recount the changes in the daily landscape over the last fifty years, evoking old and new domestic scenes through a inventory of ideas and evocations. As with all inventories, the principal way to read it is not as a piece of continuous prose (though nothing forbids this), but to look here and there, skip, zigzag like the knight on a chessboard. Everyone can choose their favorite opening: they can pick here and there, nibble at the sides or take a bite out of the middle. The alphabetical order is just a grid to guide the reader through the network of cross-references. Facts, stories, people, things, poems, quotes, books, films, sculptures, scents, tastes and games are interwoven to recount the changes under way. Each entry relates to one or more others, weaving a fabric, sometimes clear, sometimes rather enigmatic, just like the fabric of our everyday lives. So here are some fifty snippets, roughly as many as the years we are speaking of. Rapid as emails, short as an SMS, icastic as an MMS, fake-real as a reality show, harrowing as your first video game, complex as a hypertext, dense as an iPod, thin as a CD and funny as a cartoon. They form a network and resemble your new domestic landscape: the toaster speaks to the washing machine which warns the TV set not to trouble the boiler while it sends a message to the vacuum cleaner which blows on the thermostat which tells the watering can to remind the refrigerator about the shopping list which the entryphone should pay while the oven lights the candles on the low-calorie cake for your birthday, of course as it reminds the scales, which works the alarm and tells the toaster to keep mum.

A

Ambience
New perceptions, new relationships, new stories. The domestic environment and its dimensions are changing, redefined by new forms of communication. The interaction between inside and outside makes boundaries flexible. The introduction of intelligence is not alone in the machine but in the ambience, which becomes a living, active, place inhabited by invisible servants, guests, messengers. Will they learn to be as efficient and discreet as the swarms of servants below stairs in *Gosford Park*?
cf. artificial (intelligence), house this/here, Rilke/shelter

American (model)
It laid down the law for the first generation of appliances. The "American kitchen" was the dream of Italy in the boom years. White, glossy, functional. The little woman that lives there fluttered between a gleaming hob and colored formica shapes. With a narrow waist and stiletto heels, she was the angel of a hearth from which the ash had disappeared forever while the scent of the new detergents hovered in the air: the industrial tensioactives defeated Marseilles soap, definitely out of fashion until the "smell revival" of the last decade. This was the realm of the plastic Mom of Lolita, the Doris Day of the Hollywood comedies: the "exemplary wife and mother" who had endured the wartime economy and the shortages of the postwar years, freed from the slavery of domestic work, transformed into a garrulous pin-up and throwing herself into the happy consumer adventure…
cf. Mona Lisa Smile, Nabokov, Pop Art

Androids
The fantasy of the automaton, the robot, the replicant has populated the imagination at least since the mid-eighteenth century. For centuries we imagined that reproducing the human being in mechanical form was the way to lighten labor. Instead IT has reshuffled the cards and changed the rules of the game: not an android going here

One of the first Ariston Spazio gas range, 1960s.

Ariston Unibloc kitchen, 1960s.

Catalogue cover,
Ariston Unibloc, 1960s.

and there, but information traveling in a network is the main creator of change…
cf. labor, machines, biotechnological symbiont

Artificial (intelligence)
"Today Artificial Intelligence has overshadowed the construction of thinking machines and, more realistically, has set itself the objective of building portions of intelligence into complex systems."
Antonella G. Pizzaleo, *Fuzzy Logic*
Now that the sensationalist success of AI has declined, theoretical research is working more on the "cognitive sciences", while applied research prefers the path of fuzzy logic, where low definition gives a higher yield. Sensor-rich media embedded in the environment are closer to the flexible commonsense of the housewife than the implacable rigor of Hal 9000.
cf. fuzzy logic, uncertainty, smart adapter

B
Bauman
The sociologist Zygmunt Bauman sees "fluidity", "liquidity", as the metaphors most pertinent to the present phase of modernity. From standardized time to flexible time, from heavy modernity to light modernity.
Even the home, the family, identity and emotional connections form part of this agile and elusive mobility.
cf. family, fuzzy logic, home, Oldenburg, ubiquitous computing

Big Glass (The)
La Mariée mise à nu par ses célibataires, même (1915-1923) is the best-known of Duchamp's *machines célibataires* ("bachelor machines"). An enigmatic plot of allegories and symbols, beginning with the punning, ambiguous title, it is the emblem of the exchange between man and machine.
Mechanical allusions, optical illusions, mixed technological and organic forms: a hybrid morphology that produces multiple perspectives and plays with the unconscious in its rebounds between love and death.
cf. biotechnological symbiont, machine, Virilio

Biotechnological symbiont
Homo sapiens is a symbiont at increasingly high technological intensity.
"We are starting to become the constituent elements, the neurons, the organs, the cells, of a planetary creature that has developed so far by means of the typical mechanisms of every complex system...

Opposite page
Ariston Polare refrigerator, 1960s.

Ariston Combinet cooker, 1960s.

Ariston Unibloc kitchen, 1960s.

On pages 92-93
Kitchens furnished with Ariston Sistema OSA units.

This creature could become the location of a collective intelligence and perhaps of a collective consciousness, and in it a progressive confusion is being created between natural and artificial." Giuseppe O. Longo, *Il simbionte*
cf. Gibson, gruyere, complexity

Birds/chirping
"Machines are boring with their uniform constant rhythm. By using suitable applications we could build machines that make a noise resembling the chirping of birds." Bruno Munari, *Macchine aritmiche*, 1951
Something has been done along these lines: refrigerators have begun calling us if they are left open, the oven warns us the cake is baked and should be removed. Machines function silently and speak to us. But a lot still has to be done on the synesthetic plane suggested by Munari; and the magical world of the ring tones of mobile phones and the iPod opens up unsuspected possibilities…
cf. experience, Rilke/shelter, super silent

Blanchot
"Being repeats itself. This is the deep meaning of the existence of machines; but if the being were inexhaustible

excess, there would be neither repetition nor mechanical performance. So technology is penury, poverty of being become the power of mankind; a fundamental feature of Western culture." For Maurice Blanchot, repetition and scarcity are the essence of technology. But it is still so? Doesn't information produce more complex (and abundant) connections than simple repetition?
cf. biotechnological symbiont, machines, Virilio

C

Chaplin/Modern Times
Charlot, the pathetic victim of the ferocity of modern times, in an extraordinary and unsurpassed satiric representation of the conflict-ridden nodes of modernity and mechanization. The madness of the assembly line, the crazy and violent rhythms of capitalist society and alienating industrial rhythms are transfigured into the irresistible humor of the poet and tramp.
cf. Big Glass (The), machine, Uncle (My)

Complexity
"The two main components of complexity are the contemporary weaving of a number of plots and uncertainty. It is necessary to create a complex thought that uses polyvalent logics capable of creating a dialogic relationship between complexity and uncertainty. The latter cannot be eliminated and so has to be functionalized."
Edgar Morin, *Interview*
cf. fuzzy logic, Gibson, uncertainty

D

Deus ex machina
In the classic Greek theater this was the stage mechanism that enabled a deity or the hero to turn up at the end of the play or to unravel a situation that seemed to have no solution. From the ancient theater to the contemporary domestic scene, the god that appears from the machine, the good genie that magically solves problems, is a metaphor that never fails to enchant (though at times it threatens to transform itself into a *poltergeist*).
cf. gruyere, machine, smart adapter

Dialogic
Electronic dialogues, machines that converse between

them. The dialogue, as humans know all too well, is one of the few self-correcting systems of communication: it involves reciprocal adjustments, developments, compensations.
New-generation appliances work the same way and information makes it possible to create an increasingly personal itinerary.
cf. family, Margherita, smart adapter

Display
The digital window that links appliances with computer, mobile phone, etc. The term derives from the Latin displicare, to explain, but the idea of play is also echoed in the word: playful features are steadily increasing in technology…
cf. family, play, writing/graffiti

E

Experience
Key word in emotive-sensory marketing.
Home appliances do not merely perform functions, they also offer services.

Even baking becomes an experience, a cognitive advance, sensory interaction: touch control, fast chef, fast clean… Experience, Ariston oven, 2003.
cf. family, environment, labor, four minutes, thirty-three seconds

Extension (radius of)
"The radius of extension of motion increases although living space tends to shrink.
People can be everywhere without having to be present, can be informed about everything without having to leave their location". Hajo Eickhoff, *Casa*
cf. biotechnological symbiont, family, gruyere, Virilio

Different arrangements
of Ariston Unibloc kitchen,
1973.

F

Family

Once there was the family: daddy-mommy-kids (plus sometimes grandparents and other close relatives).
The home rotated around MOMMY, SHE was the pivot and the driving force of domestic activity, the bearer of the technological knowledge needed to operate the complex machinery to which she was slave and queen.
Then, of course, slowly and gradually, the family dissolved into a thousand streams: affluent or marginal singles, DINK couples (dual incomes, no kids) in various assortments, social climbing or low consumption duos, children and babysitters, chatlining grandparents and dogs, hydrangeas and parrots, but with a single constant: there was no longer one person who worked (the woman) and others who amused themselves (all the rest), but everyone does everything, alone or in company, at set hours or whenever they happen to, from love or from necessity, and—the most important point—they all have fun and no one labors because the labor is done by the band conducted by the great network…
cf. home appliance, Bauman, environment, Nabokov/Magnotact

Four minutes, thirty-three seconds

The duration of the well-known piece composed by John Cage in 1952. *4'33"* is a silent score rigorously performed by skilled instrumentalists in absolute silence. The performance is a long pause during which the public (besides possible meditations on the relativity of time, the perception of duration, etc.) discover that music can even be silence and that the sounds of life—a fit of coughing, breathing, chance noises—can become music.
The musical event is based on a network of connections whose fabric differs each time and produces a unique experience.
cf. birds/chirping, experience, Pop Art, super silent

Fuzzy logic

An apple. I eat it in a series of bites until the apple is transformed from a thing into a *no-thing*. When, at what point, does the change take place? Where is the boundary line between apple and non-apple? This is the classic example used to explain fuzzy logic to non-specialists.
It is a vague, blurred logic, corresponding to the uncertainty and indeterminacy of fluid modernity, as against the old Aristotelian logic, which was rigorous in its prin-

Ariston built-in oven, 1984.

Ariston built-in hob, 1984.

Ariston built-in dishwasher, 1984.

Kitchen with Ariston built-in appliances, 1984.

On page 98
Ariston Supermarket, 1986.

On page 100
Ariston refrigerator, freezer, oven and hob, 1997.

ciples of identity, either/or, the excluded middle. If Edgar Morin saw the introduction of fuzzy logic within the ambit of mathematical logic as an important step towards complex thought, in the ambit of everyday objects, including home appliances, fuzzy logic makes it possible to design flexible, low-cost, high-performance machines, that work by successive approximations, as humans have always done.
cf. artificial, Bauman, complexity, uncertainty

G

Game (of the glass beads)
Year 2400. Castalia is the place where of this mysterious game is played. It was created out of a combination of music and mathematics, developed in a higher order of language that interweaves sounds, colors and knowledge in harmonic compositions of extraordinary beauty. The teachers of the game want to keep it confined to Castalia, far from the disorder of the world, but the central character, Josef Knecht, wants to play it in the world...
The possible metaphors are endless, from the esoteric to the technological and allusions to the magic of computer languages. It is no accident that many glass beads travel the Web and some futuristic domotic paradises resemble Castalia... Herman Hesse, *The Glass Bead Game*
cf. Gibson, play, Virilio

Gibson
William, the inventor of cyberspace: "A consensual hallucination experienced daily by billions of legitimate users, in every nation, by children being taught mathematical concepts... A graphic representation of data abstracted from the bank of every computer in the human system. Unthinkable complexity. Lines of light ranged in the nonspace of the mind, clusters and constellations of data. Like city lights receding." This is the description of cyberspace in the famous novel *Neuromancer* of 1984, in which Gibson invented the name and imagery of the virtual worlds generated by information technology networks.
cf. complexity, gruyere, biotechnological symbiont

Gruyere
"We should devise a new type of house. In fact we have already begun to. Whole houses with roofs, walls, windows and doors exist now only in fairy tales. Material and immaterial cables have riddled them with holes almost like a piece of gruyere [...] The whole and solid house has became a ruin, and the wind of communication blows through the cracks in its walls [...] This is what the new type of home should be: a distortion in the interpersonal sphere from which relations are attracted. Such an attractive home should combine these relationships, elaborate them in the form of information, store them and transmit them. A home as creative as a node in an interpersonal network."
Vilém Flusser, *A Philosophy of Design*
cf. house, environment, Gibson, this/here, Rilke/shelter

H

Hard/soft
End of the hard era, beginning of the soft. When did it happen? The boundary is hermetic and blurred...
cf. Bauman, fuzzy logic, Hermes, Oldenburg, ubiquitous computing

Hermes
"With wings on its feet, light and aerial, capable and agile and adaptable and laid-back, he relates the gods to each other and to men, between universal laws and individual cases, between the forces of nature and the forms of culture, between all the objects in the world and between all

LURPAK
LURPAK
LATTE
FRESCO

thinking subjects." Italo Calvino, *American Lessons, Rapidity*
cf. fuzzy logic, hard/soft, speed

Home appliance
This is a machine with the function of replacing human actions (like the washing machine or dishwasher), or inventing new actions (like the refrigerator or the vacuum cleaner) in the home. It is therefore a special machine that enters the home, encroaches on our privacy and so becomes charged with stronger emotional values and symbolic meanings.
After having completely transformed the domestic scene, changed family life, altered housewifery and the female role in depth, is the traditional appliance now becoming a dinosaur?
cf. Short and Extraordinary (Stories), Mona Lisa Smile, Quim Monzò, writing/graffiti

Homo domesticus
The home is the place where man is transformed from hunter-gatherer into *homo domesticus*. Settlement favours the mental development of complex faculties like foresight and forward planning. An enclosed space, bounded by walls and a roof, modifies sensory perceptions.
"In the house man learns to listen to his inner self. Social relationships are formed at the moment when the senses become accustomed to ever smaller distances. [...] The manifold limitations of the body and the senses forms the space of the soul that unfolds in the elements that are threshold, walls and door." Hajo Eickhoff, *House*
cf. house, gruyere, Rilke/shelter

House
"The house is the incessant breath of man. Houses are dreams, thoughts and feelings expressed [...] Through the house man speaks to the world."

ARISTON

Hajo Eickhoff, *House*
cf. environment, gruyere, Rilke/shelter

I
Information
"Matter has acquired a third dimension, besides those of mass and energy: information [...] Mass no longer counts, energy still counts, but what counts most is information."
Paul Virilio
cf. dialogic, gruyere, RFID, Virilio

Interface
The surface between two spaces, the element of contact or boundary—both limit and connection—between two otherwise incommunicable entities. It is the new two-faced Janus of our age, the god that opens and closes, that unites and divides. It is the omnipresent *deus ex machina* that has overcome the incommunicability of the twentieth century: there is an interface for everything, everything is an interface, the important thing is not to lose interface!
cf. deus ex machina, dialogic, smart adapter

Intermezzo. To take our bearings
Now we're half-way through the game, let's try to sum up what has changed in the system of domestic objects. Let's take the washing machine, the mother of all home appliances. The first washing machine was worked by a handle and still required a certain amount of elbow grease on the part of the housewife. Compared with the washboard in the tub or the stone laundry steps it was already a considerable advance, but the true revolution arrived with electrical appliances: the washing machine first became electric and a generalist, then it became electronic and a specialist. Today, there's a new breakthrough, the washing machine has become information technology: no longer a self-enclosed monofunctional but an active and sensitive node in a network of highly differentiated and integrated functions. No longer one part of a system of objects, but an honorary member of a system of services. The washing machine's interlocutor has also changed over the years: in the beginning it was the housewife, her hands still chapped from doing her part; then it was the housewife of the boom years, competing with her neighbors to see who had the whitest wash (and hands, finally ladylike...); some year later it was the emancipated post-feminist woman, with a more or less significant career, who has no need to know all about textiles and stains, she just has to choose the right program. Then, finally, the washing machine's interlocutor is a computer that chats in a network with other machines, chooses the program on its own after checking the label, chooses the right time and the right detergent, the amount of water and the degree of dryness. And with a computer—through a keyboard, touch screen or a vocal receiver—everyone speaks, women and men, executives and poets, adults and children, whether present or absent, nomadic or settled. Even ultra-centenarian granddad in his wheelchair speaks through the interface: he no longer needs a flesh and blood minder and prefers to be minded by software, soft, very soft...Ready to start off again? We'll start from the fifties.

K
Kitchen
"The place I like best in this world is the kitchen..."
So begins Banana Yoshimoto's best-known story, *Kitchen*. Old kitchens and new dominate the dreams of the central character: the kitchen is the place where she imagines the warmth of the family, overcomes solitude, invents a strange family for herself. This little book became a world bestseller at the start of the last decade of a century whose privileged literary settings had been parlors and salons, boudoirs and bedrooms, living rooms and lofts. In the nineties the kitchen became the most vital and convivial center of the home. The altar of Hestia, where the rites of the raw and the cooked were celebrated, again became the most important place (naturally together with the bathroom, the other "liquid" front of the house). The new culture of eating and drinking also formed the pivot of social relationships, the place where tradition was reclaimed and the gymnasium of the new technologies, of material and immaterial pleasures, where people enjoyed, measured and displayed their quality of life.
cf. Bauman, family, house

L
Labor
A term that has disappeared from domestic terminology and been replaced by well-being, comfort, convenience, fun, excitement, experience, play, entertainment, pleasure, relaxation, amusement...
cf. experience, family, play

Lolita
At her last meeting with Humbert Humbert, a newly wed, pregnant by her Dick, the former nymphet snacks on marshmallows and crisps: she is about to leave with her husband for Alaska where they will realize their American dream. Mindful of her consumer passions, it is not difficult to imagine Lo's future kitchen, perfectly aligned with the cliché and surely repulsive for her Sugar Daddy...
Vladimir Nabokov, *Lolita*
cf. American (model), Mona Lisa Smile, Nabokov /Magnotact

M

Machine
This is the strongest myth of modernity, in continuous metamorphosis. The heroic protagonist of Futurism, it soon displayed its first disruptions, becoming increasingly radical: the Marcel Duchamp machines célibataires and Bruno Munari's "useless" machines, Jean Tinguely's "infernal" machines and the "desiring" and "immaterial" machines of Gilles Deleuze and Félix Guattari. A single network of stories woven out of different threads—paradox, play, pathos, psychoanalytic drift—questions the mythopoeic energy of the machine. Twentieth-century kinetic art became the interpreter of "total dynamism", it recounted the machine in its disrupting/disrupted relationship with humans. Today what art is capable of interpreting the new relations between machines, of representing the magical superfetation of connections generated, of relating who they are and how things are?
cf. Chaplin/Modern Times, deus ex machina, Big Glass (The), Manifesto of Machinism, play, Quim Monzò, Tinguely

Manifesto of Machinism
"The world, today is of machines.
"We live in the midst of machines, they help us do every thing, work and have fun. But what do we know about their moods, their nature, their animal shortcomings? [...] No longer oil colors but acetylene lamps, chemical reagents, chrome-plating, rust, anodic colors, thermal alterations.

Opposite page
Model of Indesit Corrida dishwasher, 1993.

Model of Indesit Corrida washing machine, 1993.

On pages 104-105
Ariston Digital Leonardo, 2000.

On pages 106-107
Andy Warhol, *Multi-Colored Retrospective*, 1979. Zurich, Bruno Bischofberger Gallery.

No longer cloth and loom but metals, plastic materials, rubber and synthetic resins.
Forms, colors, movements, noises of the mechanical world, no longer seen from outside and remade cold, but composed harmonically.
The machine of today is a monster!
The machine has to become a work of art!
We will discover the art of machines!"
From *Manifesto del macchinismo* by Bruno Munari, 1938
cf. deus ex machina, Big Glass (The), machine, Tinguely

Margherita (daisy)
Petals to be picked off one by one, a song in the seventies, a revolutionary washing machine. Margherita Dialogic is the first wholly digital appliance in the world. An intelligent machine that works on the basis of "virtual sensors" controlled by fuzzy logic. Margherita 2000. com is the first appliance that speaks to computer and mobile phone via Internet.
cf. home appliance, dialogic, family, fuzzy logic, porthole

Mona Lisa Smile
If the film was nothing exceptional, the setting was stunning: Wellesley College 1953, the educational institution for girls of good family waiting for husbands, where they could best be groomed for the "role to which they were born."
Wellesley College has a reputation: in 1941 Vladimir Nabokov, freshly arrived in the United States from Europe, was invited to give a series of lectures on literature (and he parodied the college and its characters at various times in his stories). Some years later it was also Hillary Rodham's college before she became Hillary Clinton. In the film, faces, gestures, dress, all impeccably fifties in style. Don't miss the scene where newly wed Betty proudly shows off her splendid new electric appliances...
cf. American (model), Lolita, Nabokov/Magnotact

N
Nabokov/Magnotact
"[...] When you turn a knob, your set
Will obligingly exhale

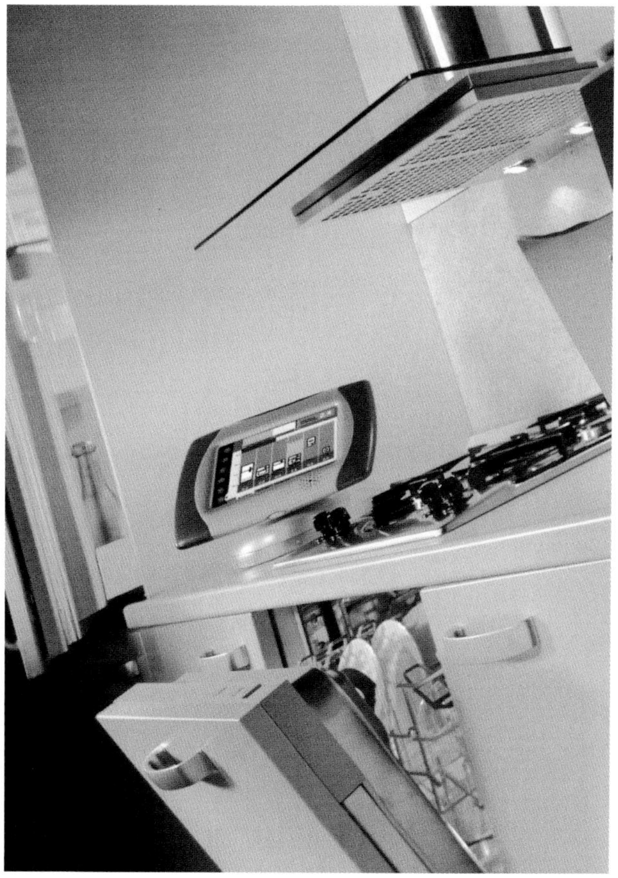

Leon@rdo

Campbell's
CONDENSED
TOMATO
SOUP

Campbell's
CONDENSED
TOMATO
SOUP
Campbell's
CONDENSED
TOMATO
SOUP

evolution
INDESIT

Forms, invisible and yet
Tangible—a world in Braille.

Think of all the things that will
Really be within your reach!
Phantom bottle, dummy pill,
Limpid limbs upon a beach.

Grouped before a Magnotact,
Clubs and families will clutch
Everywhere the same compact
Paradise (in terms of touch)…"
Vladimir Nabokov, *Voluptates Tactionum*, 1951

From Nabokov's imagination, long before the electronic glove, came a sort of TV set that produced three-dimensional objects and holographic images, but it still looked ahead to hard technology…
cf. biotechnological symbiont, hard/soft, Lolita, Virilio

O

Oldenburg
The sink and the hairdryer, the vacuum cleaner and the fan, the telephone and the water closet: clunky and hard-edged domestic objects, are transformed into soft cascading sculptures: they are Claes Oldenburg's sculptures made of soft vinyl, stuffed with kapok, spongy and lightweight. Functional elements of everyday life acquire anthropomorphic and erotic overtones. In the sixties, one of the masters of Pop Art foreshadowed the transition from clunky modernity to lightweight modernity, from hard technology to soft, from use value to the relational value of the object.
cf. Bauman, hard/soft, Pop Art

OSA (Open System Ariston)
Rigor, logic, freedom, lightness. Integration of furniture with appliances. The American model declines: an anticipation of the *zeitgeist*. Makio Hasuike, 1977-1978, awarded the Compasso d'Oro 1979
cf. American (model), home appliance, Mona Lisa Smile, Unibloc

P

Play
"These machines say strange things, speak an incomprehensible language, every so often they forget to finish what they were saying and after some seconds' silence come back and dots the i's." Bruno Munari in *Macchine aritmiche*
Munari sees play as a form of constant creative gymnastics. In machines, and more generally in modern kinetics, he grasps first the playful aspect, intuiting its scope for development, and not only on the artistic plane.
cf. labor, game (of the glass beads), machine, birds/chirping

Pop Art
Electric appliances dominate the imagery and become the alphabet of a new artistic language. Everyday objects and the consumer images of industrial society become exaggerated and serial. The domestic revolution is projected into grotesque, hypercolored fetishes.
cf. American (model), Mona Lisa Smile, Oldenburg

considers it substantive evidence of psychotic aspects of the personality.
cf. play, Margherita, uncertainty

Power modulation
This is the communication technology devised by Wrap to link appliances in a network without supplementary costs. It proposes a single shared standard for the market that would enable domotics to be more widely used. It is the challenge that undermines old market logics, by suggesting new forms of collaboration and partnership.
cf. home appliance, Gibson, smart adapter

On page 108-109
Indesit refrigerator, 2003

Indesit Evolution washing machine, 2003.

Model of Indesit Graffiti refrigerator, 2004.

Porthole
"The rule by itself is monotonous, only the case makes restless. The combination between rule and case is life is art is imagination, balance." Bruno Munari
Careful and prolonged observation of the porthole of a washing machine in motion is one of the possible situations in which best to study the interference between rule and case: in the play of colors and forms of garments, variously combined with the patterns created by the suds of the detergent, while the regular motion of the drum produces ever new and unpredictable figures. Mysterious sequences and charming to study and to ponder on at length, but don't tell the psychoanalyst: classic psychiatry

Q

Quim Monzò

Grumblings from the boiler, disobedient faucets, a drowsing coffee grinder, uncontrollable screws and springs: the rebellion of domestic objects is the theme of an amusing tale by the Catalan writer. The disruption of everyday life is still mechanical in kind and the dislocated levers are the springs of creativity, the avatar of the immaterial has not yet arrived…

cf. ambience, machine, Virilio

R

RFID

The end of the era of maltreatment. The label narrates and expresses desires, the washing machine listens and complies. RFID (Radio Frequency Identification) technology creates new dialogues and is used to create a washing machine capable of reading the smart labels on garments in the wash; the refrigerator has a display that shows the foodstuffs it contains, the optimal temperature for preserving them and their best by date; the oven automatically chooses the right cooking program.

cf. dialogic, home appliance, information

Rilke/shelter

“Almost every thing tends to a contact,
at every turn comes a murmur: remember!
[...] What original experience of ours
enables beings to understand each other?
Indifferent objects to bring us heat?
House, edge of the prairie, evening light,
suddenly acquires a human face
and to us you cling and embrace embraced.
A single space fills every being:
the inner space of the world. Silent birds
traverse us. Oh, I that wish to grow,
I look outside and in me there grows the tree.
I am in anxiety and in me the home rises.
I seek shelter and here in myself is the shelter [...]”
Rainer Maria Rilke, *Poesie sparse*

cf. house, biotechnological symbiont, birds/chirping

S

Short and Extraordinary (Stories)

“When he awoke, the dinosaur was still there.”

This is one of the electrifying *Short and Extraordinary Stories* collected by Borges and Bioy Casares in a historic anthology. Italo Calvino desired a collection of stories of a single sentence, a single line. Achille Campanile wrote highly amusing comedies in two lines…

Writers’ imaginations are shot through with brevity, swiftness, conciseness, as a metaphor of twentieth-century speed, but two centuries ago Vittorio Alfieri cultivated laconic theater.

The myth of speed (and the obsession with it) began with the industrial revolution.

cf. home appliance, speed, Virilio

Smart adapter

A brilliant cube, a global genius. Created in 2004, it is a device that makes any appliance digital: connection in a network, energy-saving, information-swapping, through power modulation… Thanks to this, perhaps there is still a place for the dinosaurs.

cf. deus ex machina, dialogic, home appliance, power modulation, Short and Extraordinary (Stories)

Indesit washing machine,
2003.

indesit
LINDA
3372054
-YOGHURT
-TEA
-H2O
-BRIOCHES
-HAMBURGER

Model of Indesit Graffiti refrigerator and frames from a commercial, 2004.

Super Silent
Murmur, lullaby, freedom. The silent washing machine, unobtrusive and energy-saving. Discreet, delicate, far-sighted. An ideal wife and mother instead of a grumbling, noisy mother-in-law …
cf. four minutes, thirty-three seconds, machine, birds/chirping

T

This/Here
"This seeks to be
engineering of sense,
tracing the architecture
of beehive or crystal.
[...] Here no work on the beams,
but their lines are arranged [...]"
Valerio Magrelli, *Ora serrata retinae*, 1980
Thought and writing, the word and its phantoms.
Do the immaterial environments of our present-future resemble the minds of poets?
cf. ambience, information, ubiquitous computing

Time (saving)
"Saving time is a good thing, because the more time we save the more we will be able to waste." Italo Calvino, *American Lessons. Rapidity*
Domestic machines have made a radical change in the time spent on domestic work. And every gap we've created in time has immediately been filled up…
cf. family, Short and Extraordinary (Stories), speed

Tinguely
Dream, irony, pathos in the "infernal machines" of Jean Tinguely. They are governed by a poetic of incongruency that continually verges on the grotesque in surprising encounters between repetition and chance nature.
His machines, like Chaplin's, "chop and process the human condition: directly, without apologetic, epic, utopian multipliers or mythic, esoteric, fetishistic demultipliers; without any more ideological screens or avant-garde masks."
Pietro Bellasi, *Il cogito del sognatore di macchine*
cf. Chaplin/Modern Times, machine, porthole

U

Ubiquitous computing
This was the theory expressed in the nineties by Mark Weiser (Xerox Parc). He definitively shelved the old concept of the technological product as a solitary self-contained object, dedicated rigidly to a single function: the new technology is liquid and flexible, pervasive.
It is incorporated in the objects that surround us, it moves, it is transformed, it is omnipresent.
The computer is everywhere; "things"—for example those in our domestic landscape—were stupid but by interfacing with a medium they have become intelligent, capable of speaking to each other, working in teams, solving problems.
cf. ambience, artificial (intelligence), Bauman, information

Uncertainty
This is the *mare magnum* in which light craft, like the boats of fuzzy logic, sail best. Sailing by sight, or rather by the senses (sensors), adapting to changes as they are gradually encountered, functionalizing uncertainty with the combined interplay of inferences and relevancies.
Uncertainty is great but—saving disasters—we move in a fairly predictable range of possibilities (above all what is involved is combining likely quantities of detergent, water of a certain hardness and the type of garments to be washed).
cf. Bauman, complexity, Margherita

Uncle (My)
Jacques Tati as Monsieur Hulot comes up against machine civilization. A continuous series of gags in his sister's ultra-technological home. This clash with an aseptic and absurd modernity celebrates the incommunicability between individual and technological artifacts.
Mon Oncle, an irresistible French film of 1958.
cf. Big Glass (The), Chaplin/Modern Times, dialogic machine

Unibloc
A single steel surface unites all the appliances. A new spatial concept, the kitchen of the future.
It appeared in a symbolic year that marked the break with tradition, a turning point in attitudes and manners: new ways of living were being born and seeking objects capable of interpreting them. Makio Hasuike, 1968.
cf. American (model), home appliance, Mona Lisa Smile, OSA

V

Virilio

Faced with the doubling of the world, a real world and a virtual one made up of electromagnetic avatars, Paul Virilio—the theorist of speed who works on the metropolitan techniques of organizing time—sounds the alarm: besides telehearing and televiewing, today we can tele-touch and telesmell (but not yet teletaste) through receptors that digitalize signals sent by our senses. The digitalization of the senses generates ontological problems whose scope is difficult to imagine.

cf. information, Nabokov/Magnotact, this/here, time

W

Writing/graffiti

Poems, recipes, drawings, love letters, shopping lists, cartoons, messages, intuitions, questions, designs...

The surface of the refrigerator is no longer an inert screen but becomes a living space, a skin that can be changed to suit one's needs, inclinations, moods. Uniqueness of style and personal gesture are combined with technological information, codified and repeatable. Indesit. Refrigerator with special finish for writing, drawing, coloring, 2004.

cf. ambience, home appliance, experience

Final curtain

Since the inner world of the home has been in constant connection with the outer world, our living habits have gradually been transformed: besides all that we have always done under its roof, today at home we can work at our occupations or take a training course, make purchases or direct the flow of our money; we can find assistance, diagnosis and treatment, visit a museum in the Antipodes or consult a book in a library abroad. And when we're not at home, we can monitor and control our domestic equipment from a distance, receive information, give orders, listen to advice from our network of helpful assistants. Climate control, safety, entertainment, consumption and other activities of various kinds can be performed not now through a variety of machines but an integrated system of services. The physical space of the home is also changing. The Functionalist model, the *machine à habiter* conceived by Le Corbusier, has done its work: different functions are performed in the same flexible space at the same elastic time. Segments of the city, fragments of the world, become present and usable, to make discoveries, enjoy experiences, interact. Indeterminacy is no longer just one margin of the interiors of the home, but becomes their principal feature: porous spaces, links, inner walls become movable and active, outer walls serviceable and interfaced, ceilings technological grids.

The programmed indeterminacy of each interior responds to the creation of structures equipped with an articulated and sensor-rich skin.

At the end of the journey, the profile of the great Nanny begins emerge. Rather than a reassuring housewife from the past, her hands white with flour, she seems to be a

ubiquitous and perhaps slightly disquieting elf.
An acrobat on the optic fibers, wide-band conjuror, she creeps into the our multifunctional palmtop and our diet menu, distributes entertainment and medicine at the right moment, attuned to our needs and our moods.
Her plans for the future include taking care of little Sonny to make him an epic robot.
For the moment, she is satisfied to touch a switch: her light caress puts out the lights, closes the shutters, switches on the burglar alarm and lowers the curtain on our alphabetic theaterette.

Works quoted in the ABC

books

Bauman, Zygmunt, *Liquid Modernity*, Cambridge: Polity Press 2000.

Bellasi, Pietro, "Il cogito del sognatore di macchine", in *Tinguely e Munari. Opere in azione*, Mazzotta, Milan 2004.

Blanchot, Maurice, cited in *Tinguely e Munari. Opere in azione*, Mazzotta, Milan 2004.

Borges, Jorge Luis, Bioy Casares, Adolfo, *Racconti brevi e straordinari*, Franco Maria Ricci, Milan 1973.

Calvino, Italo, *American Lessons. Six Memos for the Next Millennium*, Knopf Publishing Group, USA 1993 (1988).

Deleuze, Gilles, Guattari, Félix, *L'Anti-Edipo. Capitalismo e schizofrenia*, Einaudi, Turin 2002 (1972).

Eickhoff, Hajo, *Casa*, in *Cosmo, corpo, cultura. Enciclopedia antropologica*, Bruno Mondadori, Milan 2002 (1997).

Feyerabend, Paul K., *Dialogue on Method*, Reidel, Dordrecht 1979

Flusser, Vilém, *Filosofia del design*, Bruno Mondadori, Milan, 2003 (2001).

Gibson, William, *Neuromante*, Mondadori, Milan 2003 (1984).

Hesse, Herman, *The Glass Bead Game*, Picador USA, 2002 (1943).

Longo, Giuseppe O., *Il simbionte. Prove di umanità futura*, Meltemi, Rome 2003.

Magrelli, Valerio, *Ora serrata retinae*, in *Poesie*, Einaudi, Turin 1996 (1980).

Morin, Edgar, *Interview*, in A.G. Pizzaleo, *Fuzzy Logic*, Castelvecchi, Rome 2000.

Monzo Quim Monzò, *Thomson, Braun, Corberò, Philishave*, in *Olivetti, Moulinex, Caffoteaux et Maury*, Marcos y Marcos, Milan 1993 (1980).

Munari, Bruno, "Macchine aritmiche", in *Catalogo Cosmit*, Milan, 1999 (1951).

Munari, Bruno, "Manifesto del macchinismo", in *Tinguely e Munari. Opere in azione*, Mazzotta 2004 (1938).

Nabokov, Vladimir, *Lolita*, Adelphi, Turin 1993 (1955).

Nabokov, Vladimir, *Voluptates Tactionum* (1951), in Riga 16, Marcos y Marcos, Milan 1999.

Pizzaleo, Antonella G., *Fuzzy Logic*, Castelvecchi, Rome 2000.

Rilke, Rainer Maria, *Poesie (1908-1926)*, Einaudi, Turin 1995.

Virilio, Paul, *Il futuro nello spazio 'stereoreale'*, interview in *European IT Forum*, 1999.

Virilio, Paul, *La velocità assoluta*, interview in *European IT Forum*, 1995.

Yoshimoto, Banana, *Kitchen*, Feltrinelli, Milan 1992 (1991).

films

2001: A Space Odissey, Stanley Kubrick, Great Britain 1968.

Gosford Park, Robert Altman, USA 2001.

Modern Times, Charlie Chaplin, USA 1936.

Mona Lisa Smile, Mike Newell, USA 2003.

Mon Oncle, Jacques Tati, France 1958.

music

Cage, John, *4'33"*, 1952.

Cocciante, Riccardo, *Margherita*, 1976.

sculptures

Duchamp, Marcel, *The Big Glass*, 1915-1923.

Munari, Bruno, *Macchine inutili*, *Macchine aritmiche*, from 1930.

Oldenburg, Claes, *Softs*, from 1962.

Tinguely, Jean, *Sculture meta-meccaniche*, from 1954.

PACE
NON
à la
guerr
NON
ON

Communication

Beba Marsano

No sex, please, we're Italian. And don't touch Mom. Above all in the kitchen. Even if she's young and pretty, she's not allowed to make your spine tingle or spark fantasies. Her job is to make sure the joint doesn't burn and the linen isn't frayed. Her role is, well, almost managerial: to run the family without giving anyone sleepless nights. In the fifties and sixties, the woman was queen of hearth and home, a faithful wife and loving mom, and it was precisely this efficient and reassuring image that took possession of the advertising for kitchen ranges, washing machines and fridges, "white appliances", so called to distinguish them from the "brown" audio and video devices for the reproduction of sound and images.
The equation between woman and the home was imprinted in the collective imagination by television, an escapist yet informative medium that rapidly revolutionized the lives of Italians by creating new behavior patterns and presenting new consumer goods. Italian television was launched in 1954 and in the same year the language of advertising began to acquire some of its present style and impact. With television came commercials. They took over the function previously performed by posters and billboards (which now began to lose their independence by quoting or echoing the story lines created for television and the press). At first commercials lacked their later effectiveness, because only the well-to-do could afford a TV set. (Popular programs like the Italian version of Double or Quit? used to be screened at cinemas, showing how few people had TV sets.) Moreover, in the early years, programs were limited to just a few hours a day.
In those years consumption was still fuelled by utilitarian products, directed at improving living conditions. But in the eighties home appliances became charged with hedonistic values. In the nineties their functioning became increasingly sophisticated and then, surprisingly, at the start of the new millennium they were stripped of all their high-tech trimmings and presented like playthings.
We will be seeing how advertising kept reinventing their image.

The '60s: a comeback. The housewife triumphant
Four women in an interior. Young, attractive, elegant, ambitious. Bonded in a female friendship that becomes speckled with envy as soon as one manages to possess something that the others don't have. In this case, an electric cooker. "I don't want to brag, but from tomorrow I'll have my magnificent Ariston cooker too!" all three exclaim with one voice veined with venom and malice in the refrain of a cycle of six commercials that the firm of Fabriano screened in 1966. This was part of what then was—and would remain to the end—one of the best loved and most widely watched shows: *Carosello*. In the sixties the spread of home appliances was still confined to transforming cookers, washing machines and refrigerators into objects to be shown off proudly, just as much status symbols as fur coats and cars. Despite the net improvement in average family income, for one population group "white goods" were still an unattainable desire, a mirage. Significantly in 1965 Ariston launched a famous competition actually called "Your Dream Kitchen" in collaboration with the weekly magazine "Amica."
In an Italy where just a few years earlier it was quite natural for linen to be washed by hand, where the cooker

burnt wood or coal and food was kept in an icebox, technology in the service of the housewife appeared quite unexpected, astonishing and rather wonderful.

Advertising, as the privileged mirror of social life, absorbed and recorded this reaction. Note the expression of amazement and incredulity of two women of different generations (mother and daughter or perhaps mother-in-law and daughter-in-law the first Ariston commercial made for the launch of the Spazio 400G cooker (1965): as they look inside the electric oven a window opens on boundless spaces, a future rich in promise. Or the smile of pleasure that lights up the face of Olga Villi when, three years later, she presented a refrigerator neatly packed with all the bounty of affluence. In its commercials Ariston liked to feature the stars of the small screen (from the actress Olga Villi to the soubrette Marisa Del Frate and the announcer Maria Giovanna Elmi, a blond fairy character adored by children). They were examples of women who, thanks to their home appliances, effortlessly combined emancipation and tradition, having careers without neglecting their institutional and domestic roles as wives and mothers.

The language of the early TV commercials (as on billboards and advertisements in the press) was as simple and direct as possible, shunning all ambiguityies in a country where educational levels were still poor and illiteracy clung tenaciously to its strongholds, despite the efforts of television itself to teach reading and writing.

On page 122
Advertising campaign
for Indesit Graffiti
refrigerator, 2004.

Unibloc catalogue, 1966.

Poster for "Carosello", 1966.

Posters for advertisement,
1960s.

una signora cucina!
Ariston
DA QUANDO
L'ABBIAMO
MESSO AL MURO
E' DIVENTATO
F A M O S O.
L'AFFISSIONE. I MURI FANNO STORIA.
A CURA DELL'AAPI.

First Ariston commercial for the Spazio 400G kitchen, 1965.

The scene was generally set in the kitchen, where the home appliances advertised occupied pride of place right in the middle. At their side, testifying to some quality of the product, there was always a woman (more rarely two) sweetly smiling, reassuring, and with a subtly confiding look, suggesting not mere advice but an authentic, precious hint offered confidentially from woman to woman. In fact right down to the 1980s this kind of woman remained the privileged and almost exclusive receiver of the messages of the advertisements for household appliances.
In the years leading up to the outburst of feminism, the market already accorded women a greater commercial importance and an undeniable freedom of choice, though it saw purchases as limited to themselves and their families. Women were the largest and almost the only pool of users of home appliances: they were not only responsible for the choice of product but they were in charge of the domestic budget. This was also, essentially, one of the effects of the slow process of gaining independence that culminated in 1946 with the law enabling women to vote in the elections to the Constituent Assembly on June 2 and the referendum that nine days later decreed the end of the monarchy and the birth of the republic.
The shrewdness of advertising knew no bounds. Even in a purely technical area like electrical appliances, in advertising the message did not simply appeal to the practical sense of the housewife. It sought to touch her vanity, where every woman is most vulnerable. Home appliances not only improved the quality of life by freeing women from the alienating routine of domestic chores, they did much more. They changed the measure of time. While the machine worked for her, a woman could gain hours of

Catalogue of Ariston refrigerators with Olga Villi on the cover, 1968.

Ariston refrigerator commercial with Olga Villi, 1968.

Ariston commercial
with Maria Giovanna Elmi,
1971.

Commercial for Ariston Unibloc featuring the couple Giuffè and Schiaffino, 1976.

Commercial for Sistema OSA, 1978.

freedom and spend it all on caring for herself. So home appliances were, in the first place, a form of investment in herself: this was the ultimate meaning of the message, conveyed above all on TV, which was then emerging as the most powerful tool of persuasion. By virtue of technology, a housewife will no longer appear before her husband and children tired, unkempt, and poorly groomed because she has washed mountains of shirts, stockings and underpants by hand and slaved over a hot stove for hours. Thanks to Ariston, even the most modest housewife out in the provinces will be able to look like the happy woman in the advertisements, a domestic starlet always well groomed, a smile on her lips and her hair perfectly styled, without a single drop of sweat jeopardizing the hairdresser's art. "La cocina italiana Ariston! que maravilla!" exclaimed a poster obviously intended for the Spanish market in 1968. It showed a young woman with a charming chef's hat set with flirtatious indifference on the square-cut hairstyle that frames a satisfied, rested and perfectly made-up face.
The euphoria of the boom years of fifties Italy had already buried the troubles of the war and fed further growth in the following decade. Now it was accompanied by the grand illusion of a carefree life of affluence and happiness, in which personal beauty and the possibility of caring for it meant that the days of privation and hardship were gone for good.

The '70s: the wild ones but longing for home
From the carefree, illusory, heady days of happiness to the endless nightmare of the "Years of Lead". From north to south, Italy in the seventies was scourged by terrorism (unleashed by the Red Brigades and many other groups with similar names). A chain of massacres, assassinations and the "strategy of tension" tore shreds in the social fabric and inflicted deep wounds in individual consciences. To all this, still on the wave of '68, there were added violent collective tensions that exploded almost daily into episodes of urban guerrilla warfare, with policemen-warriors armed with helmets and tear gas lined up against youngsters, above all students, with their faces masked by balaclavas.

Commercial for Ariston Margherita washing machine, 1985.

Commercial for Ariston Supermarket refrigerator, 1985.

Commercial for Ariston Margherita washing machine, 1987.

TV sales pitch for Ariston Supermarket refrigerator, 1986.

Advertising campaign for Ariston Supermarket refrigerator.

The decade of the seventies "was marked by acute social conflict, terrorism and a policy of economic austerity caused by the international oil crisis. This was the most highly politicized period, but it also saw a retreat towards privacy." So wrote Ugo Volli in his introductory notes to the advertising exhibition organized that year at Rivoli as one of the events marking the fiftieth anniversary of the RAI, Italian public service broadcaster[1]. In the light of this statement, there seems to be a significant connection between this retreat behind the domestic walls and the dizzy increase in sales of home appliances recorded in the same years. In a survey published in the *Enciclopedia Europea* under the entry "Elettrodomestici", we read that in the two-year period 1975-1976 their level of use in Italian families was one of the highest in Europe: 93% of the population had a refrigerator (against 88% in Britain), 90% an electric cooker and 75% a washing machine (compared, for example, with 66% in France)[2].

One again Ariston interpreted to perfection the spirit of the times and responded to this widespread feeling of bewilderment that affected both society and the individual, a need for protection and security that only the family and the home could satisfy. In its general catalog for 1972, the Fabriano company conveyed this need for warmth through its home appliances, renaming the cluster of refrigerators, cookers, dishwashers and washing machines "the faithful" on the cover, and presenting in the logo, a sort of ideal embrace, with one clinging to the other against the body of a dog (a charming dachshund), the supreme symbol of fidelity, sincerity, companionship and affection. In this period of uncertainty, Ariston reassured its public it would never betray them. At the same time the home-appliance concept evolved together with the approach to advertising. An illuminating example was Unibloc, the system that integrated home appliances with the kitchen furniture in a continuous steel work surface, marketed by Ariston first in the second half of the sixties and then even more incisively in the following decade.

Poster for Combinato Nabila 2, 1988.

The name of the product, perhaps too technical-sounding, might have intimidated the less sophisticated housewives. So the firm had recourse to a simple, but effective image. For the cover of its Unibloc 1966 catalogue, Ariston chose two children (the same that had appeared on the brochure of its modular kitchen units the previous year), who were shown playing with the integrated kitchen elements. The message? Unibloc offers furnishing solutions that combine pleasure with the simplicity of a child's game. Ten years later, again for the same product, the idea of play was kept but modernized. The 1976 advertisement—featuring Rosanna Schiaffino and Carlo Giuffré—a used a very different language from the kind used by the company beforehand.
A modern, functional language, based on a rational, geometrical structure with vaguely neo-constructivist overtones, markedly different from the cloying narrative realism in sparkling comedy style of the sixties ads. Here the principle underlying Unibloc was expressed visually by using the expedient of a construction toy: a gigantic hand is selecting a series of cubes and rectangles of different sizes and fitting them inside a transparent box that evokes a wall. A woman can build her own kitchen to suit herself, just as a child, following the whim of the moment, plays with building blocks.
In the sixties and in the seventies Ariston entrusted its image to the bare force of its products, unlike other brands (not only in the field of home appliances) which, thanks to the brilliance of graphic designers and famous photographers, invented characters who appeared in

INDESIT TORNA IN TV. CON CAROL ALT.

Indesit commercial featuring Carol Alt, 1990.

advertisements and were identified with a product to the point where they actually replaced it (like Paulista Coffee's Caballero, drawn by Armando Testa).

Others coined slogans that became part of everyday speech (like the unforgettable Erberto Carboni's "È sempre l'ora dei Pavesini" (It's always Pavesini time) or devised sophisticated, spectacular, provocative image campaigns that were indelibly impressed in the public's memory. One example was "Jesus Jeans", a campaign devised by Oliviero Toscani, which caused such a scandal that the posters were impounded and there was a trial for blasphemy[3]. In this decade public service television expanded: a second channel was added, then a third. The result was pluralism not only in information and entertainment but a clear differentiation between categories of viewers or, in the language of marketing, of its target groups. In the early seventies color television was developed, announced for the Olympic Games of 1972 but postponed as part of the austerity package in response to the international oil crisis.

The '80s: Nothing Underneath

"Years full of advertising" Raf called them in a tuneful and far from foolish song titled Co*sa resterà di questi anni ottanta* (What's Left of the Eighties), presented with some success at the Sanremo Song Festival in 1990. *Sono solo canzonette* (Just Songs), someone else might have answered[4], but Raf got it right. Advertising, with its fake myths and fake gods, was the most complete and striking expression of this empty and fragile decade, frivolous and superficial, anxiously devoted to the religion of the image, of surfaces, of appearances. (It is curious to note that in 1983 Ariston launched the Mosaico kitchen, a system that embedded the home appliances in the kitchen furniture with an integrated formula that was more aesthetic than technical). *Nothing Underneath*, to quote the title of a film by Carlo Vanzina that was a box-office success in the same period (though the critics panned it).

The eighties bred yuppies, young people without commitments but ambitious, and stylish, driven only by the val-

Commercial and drawings for the "Why Ariston?" campaign, 1992-1995.

Opposite page
Ariston commercial by Wim Wenders, 1992.

ues of money and success.It was in this precise context that Italy became the deputed center of luxury, with the miracle of Italy style and the international success of Italian designers (from Armani to Versace, Missoni, Moschino, Biagiotti and Krizia), and at the same time the theater of the apotheosis of the consumer society, following the emergence of private or commercial TV channels. Their purpose, it was acutely observed, was not "to produce programs, as it is or should be for public service television, but to create, through specially tailored programs, the body of viewers/consumers required by the market of commercial investors"[5].

In these years there was only one commandment: sell, make money, stimulate consumption by creating new needs. Never as at this time had the market fostered a more technological and consumer-minded "material culture" as the sociologists called it, instilling in the public a belief that it was prey to an endless series of unsatisfied needs. The product—as can be deduced by analyzing the campaigns of Ariston, among others—became symbolically the protagonist of advertising, occupying the whole visual field (of the commercial or poster) and pushing human figures out of the picture. Slogans were simplified to the utmost to avoid creating a distraction from the name of the product.

In the sphere of production, the eighties were happy years for the company based in the Marche, which turned out products that enjoyed great commercial success. Its products were given simple names, easy to remember. Who doesn't remember the Margherita washing machine, the first one with a "woman's name", accompanied by the styl-

www.b

ARISTON
La casa vale di piu'

Indesit "Platform"
commercial, 1996.

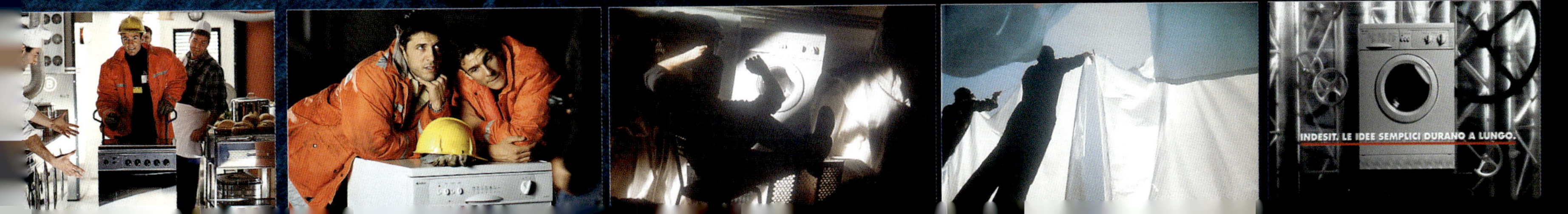
INDESIT. LE IDEE SEMPLICI DURANO A LUNGO.

Ariston logo,
2003.

ized drawing of a flower and launched on the market in 1983? The slogan "Porta a casa una Margherita" (Take home a Margherita) was the most logical, almost banal, invitation that can be imagined. This was followed closely by the Quattro Stagioni refrigerator, with its innovative system of pouched containers set in the door (an idea taken up again and perfected some years later in the new Supermarket model) and then the Sette Cuochi cooker, whose name referred to the seven forms of cooking made possible by the functioning of the oven.
Each of these products was marketed with a visual language that varied little. The object, on a monumental scale, emerges almost forcefully from a uniform black ground, and with its enameled whiteness is imprinted indelibly, more by a law of physics than marketing, on the retina and the mind of the consumer. The image—an invitation to covet, acquire, possess—can be interpreted as the visual transcription of a dream that, if left unsatisfied, will grow till it becomes a need, a fixation, an obsession. Apart from its apparent innocence, this type of advertising conceals an unexpectedly aggressive force of persuasion. Faced with such a simple message, the observer is not primed to adopt some defensive strategy and, for this reason, the impulse to purchase becomes charged with an overpowering and almost irresistible force. The message that is presented in the guise of kindly advice is actually no other than an urgent, almost compelling exhortation to buy.
In the consumer bulimia that, like a real compulsive obsession, ran all through the eighties, the product acquired stellar status, a star of the first magnitude of the small screen. It took a leading role in a new form of advertising: TV marketing. A TV slot, lasting a few minutes, embedded in a prime TV program, which any business could buy and use to pitch its products to the audience. Ariston responded to the potential offered by this new advertising tool. In 1986 the launch of its Supermarket refrigerator included a cycle of TV promotions in a very popular big-audience program titled *Il pranzo è servito*, presented at midday (mostly watched by housewives) by an immensely popular, sardonically smiling presenter called Corrado. Two years later, in 1988, a quite different personality lent her face to Indesit's Missy washing machine: Carol Alt, the American model and actress most in vogue at the time, an icon of yuppidom triumphant, a symbol of beauty and success and the lead the previous year in another film by Vanzina, *Via Montenapoleone*, which epitomized all the vacuity summed up in the slogan for a liqueur: "Milano da bere", the epicenter of the mores and modes of the eighties.

The '90: internet, the decline of the housewife and the rise of the househusband

In the last decade of the century the new technologies, electronic and digital, were perfected, expanded and triumphed across the world (and gave birth to globalization). The discovery of their endless potential and applications meant their use was not confined to business, but extended into the private sphere, including leisure and home. The concept and the interior of the home changed radically and it was a change that first affected all those appliances hitherto designed to meet the normal, daily requirements of cooking and preserving foodstuffs or washing dishes and clothes. Home appliances were no longer conceived solely and exclusively for woman but to be used by every individual member of the family. This was not just a clever piece of marketing, but the result of most important change to come over the Italian family at the end of the millennium.
Women (not only in quest of self-fulfillment but to make a concrete contribution to the family budget) no longer worked at home but outside it and the figure of the house-

Ariston dishwasher commercial "Movie", 2003.

www.indesit.com
Νέα πλυντήρια Indesit.
Εξολοθρευτές λεκέδων.
Σήμερα, η τέλεια πλύση είναι... παιχνιδάκι: χάρη στα νέα πλυντήρια Indesit, που είναι εξοπλισμένα με πρωτοποριακά προγράμματα και λειτουργίες. Όπως αυτή που ενεργοποιεί το πλήκτρο Stain Erase για να εξαφανίσετε, με μία μόνο πλύση, ακόμα και τους πιο σκληρούς λεκέδες. Αν, μάλιστα, ενεργοποιήσετε και τη λειτουργία Easy Iron, έχετε ολοκάθαρα ρούχα με λιγότερες ζάρες. Κι αυτό βέβαια σημαίνει καλύτερο και πιο γρήγορο σιδέρωμα! Όσο για τα πιο ευαίσθητα ρούχα, τώρα δεν χρειάζεται να τα πλένετε στο χέρι. Το ειδικό πρόγραμμα Hand Wash προστατεύει την απαλότητά τους, και χαρίζει σ' εσάς περισσότερο ελεύθερο χρόνο για να διασκεδάζετε!
INDESIT
We work, you play.

www.indesit.com
New Indesit cookers.
Double Oven.
The new Indesit double oven gives you the flexibility to cook two different dishes at the same time. The top oven has an air flow cooling fan and twin conventional grill, while the main oven has an electronic timer and quick defrost function - the perfect ingredient for any kitchen. For a full range brochure please call 08700 104309.
INDESIT
We work, you play.

New Indesit cookers.
Perfect for cuddling
your business.
Sleep tight with Indesit's new cookers. Attractive, modern and efficient, Indesit's new cookers will charm your customers with their rounded shapes, new design, extraordinary practicality, excellent quality/price ratio and - above all - with their numerous new functions.
Multi chef: five programme oven, ideal for all dishes.
Fast flame: extra powerful hob for reducing cooking times by up to 10%.
Safe flame: this system immediately blocks any accidental gas leakage from the cooktop and from the oven.
INDESIT
We work, you play.
www.indesit.com

Indesit international advertising campaign "We work you play", 2000-2001.

Uox
Think

Indesit logo,
1998.

wife as the queen of hearth and home vanished for good; housework was shared by the couple and the roles were often reversed: men—partly from necessity and partly for pleasure—entered the kitchen, donned aprons and did the cooking. Endless advertisements in the nineties featured househusbands! The independence of single individuals—never forget the rising numbers of singles—increased at the same time as there was a change in relations between the sexes and way they related to the spaces of the home. In short everyone had to be able to look after himself/herself as well as others, and this meant learning to use the knobs and buttons on cookers, dishwashers and washing machines simply and efficiently.

In the mid-nineties, Ariston applied the Dialogic system to all its home appliances. This was entirely based on simplified digital technology, which has numerous analogies with human behavior. By contrast the forms were softened, so users would feel that modern technology was a friendly business. If refrigerators, cookers, dishwashers and washing machines are activated inside by a series of extremely sophisticated devices, this makes using them simpler. And the point of the advertisements was to make this clear.

In 1995 the Margherita Dialogic washing machine was presented as the world's first fully digital home appliance, capable of generating "virtual sensors" so as to confer intelligence on its functions. Because this was such an exceptional product, as part of the trend towards hiring big-name film directors to do commercials, for the launch Ariston called in a world-famous director: Wim Wenders, fresh from his success in *Faraway, So Close!* and that same year at the apex of his fame for the almost simultaneous release of *Lisbon Story* and *Beyond the Clouds*. For Margherita Dialogic, the German filmmaker chose to redeem its system of artificial intelligence from abstraction by giving it a face: a bearded deity that seems to have just descended from Olympus, as a *deus ex machina* of the classical theater can solve all problems, in this case by just pressing a button. And, as the slogan says, thanks to this miraculous presence, "The home is worth more."

In 1998 to affirm the concept that Ariston appliances were as simple as ever to use, in a commercial for the new Aristella dishwasher there appeared the hand not of a god but a woman. This was also the year when Ariston Digital was perfected: the first system of home appliances linked in a network, designed to communicate with a digital meter so as to exchange information through Internet. In 1999 it was the turn of Margherita2000 (the name projecting into the new millennium). This was the first washing machine ever designed to dialogue with mobile phone and computer through the Web. To showcase this new class of products embodying an extremely sophisticated technology, advertising used a new language. With its subtle interplay of dissolves and superimposed images, the commercials for the range of Ariston Digital services, for example, seemed to borrow certain expressive modes from electronic art and videoart. By contrast, for the commercial promoting the Margherita2000 washing machine,

NON AVEVO MAI PENSATO DI FARLO SUL FRIGORIFERO.
NUOVO FRIGORIFERO GRAFFITI. PER SCRIVERE, DISEGNARE E CANCELLARE QUELLO CHE VUOI.
INDESIT
We work, you play.
www.indesit.com
LE HO SCRITTO UNA LETTERA BELLISSIMA. PURTROPPO NON POSSO SPEDIRGLIELA.
NUOVO FRIGORIFERO GRAFFITI. PER SCRIVERE, DISEGNARE E CANCELLARE QUELLO CHE VUOI.
INDESIT
We work, you play.
www.indesit.com

Advertising campaign for
Indesit Graffiti refrigerator,
2004.

the Fabriano firm took a step back. It returned to a straightforward, direct, traditional approach so as not to lose touch with the older age group among its customers. The scene is set inside a store selling electrical appliances; the washing machine is enthroned in the foreground, done up like a gift in brightly colored ribbon between a pleased vendor and a smiling, satisfied buyer. "Your next wish" the slogan proclaims. In short, tradition and a rather restrained avantgarde approach were the salient features of advertising in this period for Ariston, one brand in the stable of the Merloni Elettrodomestici group, which in the same decade also launched Indesit, a company it had taken over at the end of eighties.
While for Ariston the target has always been the family, Indesit aims at a new and younger class of consumers, the result of a steadily spreading social phenomenon that is extremely attractive to the market: singles. A man or woman living on their own wants home appliances that make domestic chores as easy as possible, leaving them time for leisure and enjoyment. For this reason Indesit produces reliable appliances that are even simpler in design and easy to use. The slogan "We work you play" sums up the brand's philosophy, just as "Time Together" sums up Ariston's.
Indesit presents itself as the leading maker of home appliances for young Europe with a series of dynamic advertising campaigns (by Ogilvy & Mather), whose objective is to efface the cultural differences between the different countries by displaying healthy, lively young people with shared interests at every latitude. The language used for the commercials and hoardings focuses on getting together, having fun, the joy of living and they show groups of young people happy to be together and engaged in their favorite sports and pastimes, while Indesit technology saves them from doing the most thankless chores around the home, giving them plenty of leisure.
The brand's philosophy puts consumers right at the center of its thinking, with the objective of giving them simple, reliable home appliances. Simplicity is the other key concept at the heart of Indesit's advertising. It has launched a series of campaigns aimed not at people but at the product. "Simple ideas that last" is the slogan accompanying the new line of appliances (designed by the skilled hand of Giorgetto Giugiaro) to convey the essential values of plain efficiency, reliability and quality.

The 2000s: technology is spectacle

The irresistible advance of technology offers manufacturers the opportunity to produce appliances for the most advanced homes, increasingly intelligent, increasingly interactive. The start of the new millennium is characterized by the adoption of equipment and devices directed on the one hand to saving energy and on the other to substantially making electrics safer.
Ariston has collected a series of firsts in this field as well. In 2002 Pay-X-Use made its appearance. It is now possible to use a washing machine without having to buy it, on the basis of a contract with rates based on use. The following year RFID (Radio Frequency Identification) appeared. This system enables the washing machine to read the tags with washing instructions on clothes, allows the fridge to indicate the foodstuffs it contains and display the optimal temperature for keeping them and best-by dates, and the oven to automatically select the most suitable cooking

Advertising campaign for
Indesit Graffiti refrigerator,
2004.

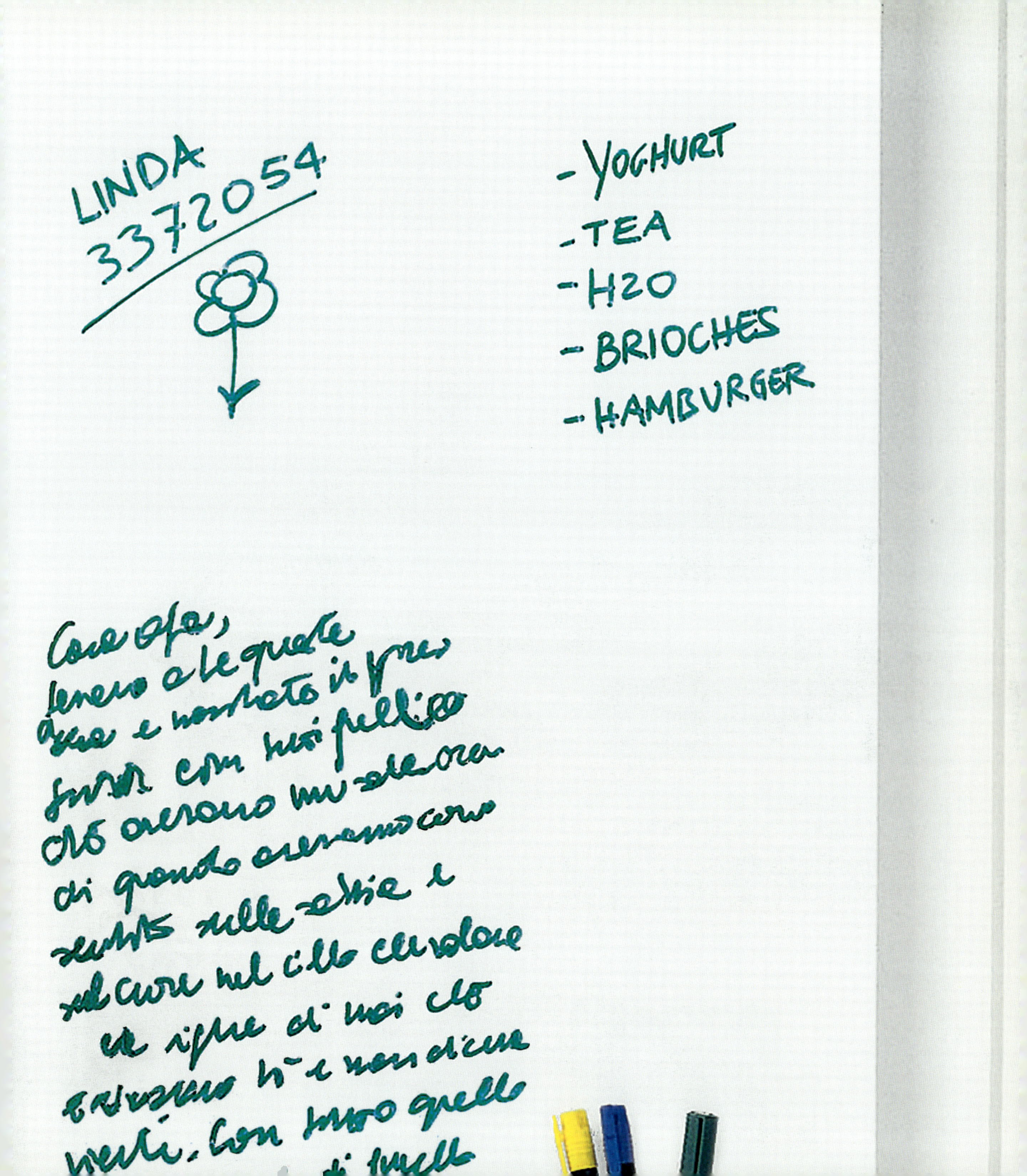
LINDA
3372054
- YOGHURT
- TEA
- H2O
- BRIOCHES
- HAMBURGER

Ariston Extendia
washing machine
and refrigerator, prototypes.

ARISTON

program. In 2004 SuperSilent was launched: this completely silent washing machine is fitted with a three-phase motor controlled by technology from Texas Istruments and can be used at night, when electricity costs are lower.
As in the previous decade, the target group for new products continues to be the whole family, including teenagers and children. In marketing its appliances, Ariston takes them into account. At the start of the 2000s, the company based in Fabriano radically updated its image and appliances, making the bravest and most revolutionary advertisements in its whole history.
In 2003, Ariston chose irony for the first time and borrowed its slogan from the cinema. The comment on a dishwasher packed with pots, lids and dirty dishes was an amusing pun on the Italian-version title of Disney's cartoon *Carica dei 101* (101 Dalmations). Another scene was set around the same dishwasher, with the whole family waiting impatiently for the cycle to end; here the allusion was to *The Dirty Dozen*, perhaps Robert Aldrich's best-known masterpiece.
Everything changed in the commercial for Ariston's New Brand Image, again made in 2003. Forgetting spectacle and the big screen, the camera eye returned to the affectionate intimacy of the home. It is evening and, framed in a living room, a family indulges in the relaxation and joy of being together at the end of the day. Mommy, Daddy and the two kids can be serenely together because someone else is dealing with the chores.
"Ariston is home" is the message with which the firm clinches the objective attained: an integrated system of home appliances, within the span of half a century, has not only completely transformed everyday habits but improved the quality and style of life, of women above all, but also of the whole family.
A philosophy summed up in the formula "Method and Freedom", coined by the Japanese designer Mako Asuike, who is responsible for aesthetic coordination at Ariston.
A playful sense of freedom, of fun, of entertainment, accompanies the most recent Indesit product, the line of Graffiti refrigerators (2004), featuring a simple but completely original idea: the door is like a blackboard, on which you can write (the shopping list, for example), draw, jot down notes, leave messages.
For the first time—as the commercial promoting it shows to perfection—the relation with home appliances is charged with a confidential, friendly, playful value, which is part of the brand's DNA. In 1999 Ariston launched its "Decorated Refrigerators", designed to give an essentially functional product a completely new look by turning it into part of the home *décor*. But the Graffiti refrigerator is perhaps the most revolutionary innovation ever undergone by the refrigerator which, in substance, has hardly altered its cupboard form, except "on the inside, from the split into freezer/refrigerator to the more refined divisions of its compartments to suit the types of food"[6]. The Graffiti Refrigerator represents a new way of communicating in the kitchen, functional and playful, with a surface where you can write and receive messages of all kinds: from the shopping list to a phone number, a quick-and-easy recipe or a love message.

[1] *Schermo delle mie brame-Come la pubblicità ha cambiato la vita degli italiani (1954-2004)*, edited by Ugo Volli, Castello di Rivoli (Turin), 7 July-12 September 2004.

[2] The same survey showed that demand was fuelled first by replacement purchases (almost 60% of annual sales), followed by first-time buyers for new families and double purchases for the same or a second home.

[3] Toscani played on ambiguity and used a wicked double entendre in this campaign studded with erotic-evangelical messages. The first poster was dominated by an irresistible pair of female buttocks with the message spread across them in big lettering: "If you love me follow me." Another poster had the slogan: "You will have no other jeans before me." It had a photo of a pair of jeans with the zip open suggestively on the crotch.

[4] The reference is to the title of a hit by Edoardo Bennato.

[5] *Novecento-Arte e storia in Italia*, AA. VV., Skira, Milan 2000, page 538.

[6] XX *Secolo-Design*, Decio Giulio Riccardo Carugati, Electa, Milan 2003, page 358.

INDESIT

Photographic Credits

Archivio Mondadori Electa, Milan
AKG Images, Berlin
Archivio Indesit Company
Corbis/Contrasto, Milan
Henri Cartier-Bresson/Magnum Photos/Contrasto, Milan
Martine Franck/Magnum Photos/Contrasto, Milan
ICP, Milan
Publifoto/Olycom, Milan

Thanks to
Makio Hasuike and Giorgetto Giugiaro for the sketches kindly supplied
Opus – Fabriano for iconographic advice
Studio di Architettura Iosa Ghini

This volume was printed for Mondadori Electa S.p.A.
at Martellago Mondadori Printing S.p.A.,
via Castellana 98, Martellago (Venice), in the year 2004